Primate Atherosclerosis

Monographs on Atherosclerosis

Vol. 7

Editors
DAVID KRITCHEVSKY, Philadelphia, Pa.
O.J. POLLAK, Dover, Del.
HENRY S. SIMMS, Rockleigh, N.J.

S. Karger · Basel · München · Paris · London · New York · Sydney

Primate Atherosclerosis

G. A. GRESHAM
Professor of Morbid Anatomy and Histopathology, University of Cambridge
Fellow of Jesus College, Cambridge

With 6 figures, 1976

S. Karger · Basel · München · Paris · London · New York · Sydney

Monographs on Atherosclerosis

Previously published
Vol. 1–3: please ask for details
Vol. 4 J.E. KIRK (St. Louis, Mo.): Coenzyme Contents of Arterial Tissue
X + 92 p., 44 tab., 1974.
ISBN 3-8055-1670-3
Vol. 5 ROGER W. ROBINSON; IVAN N. LIKAR and LYDIA J. LIKAR (Worcester, Mass.):
Glycosaminoglycans and Arterial Disease
VIII + 136 p., 8 fig., 11 tab., 1975.
ISBN 3-8055-2089-1
Vol. 6 WILLIAM T. BEHER (Detroit, Mich.): Bile Acids. Chemistry and Physiology of Bile Acids and their Influence on Atherosclerosis
XIV + 225 p., 11 fig., 9 tab., 1976.
ISBN 3-8055-2242-8

Cataloging in Publication
Gresham, Geoffrey Austin
Primate Atherosclerosis
G. A. Gresham. – Basel; New York: Karger 1976.
(Monographs on atherosclerosis; v. 7)
1. Arteriosclerosis 2. Primates 3. Research
I. Title II. Series
W1 M0569T v.7/WG 550 G831a
ISBN 3-8055-2270-3

All rights, including that of translation into other languages, reserved.
Photomechanic reproduction (photocopy, microcopy) of this book or parts of it without special permission of the publishers is prohibited.

© Copyright 1976 by S. Karger AG, Basel
Printed in Switzerland by Buchdruckerei Lüdin AG, Liestal
ISBN 3-8055-2270-3

Contents

Acknowledgements	VI
Preface	VII

Chapter 1. Atherosclerosis in Man 1
    References 5

Chapter 2. Structure and Function of the Normal Vasculature .. 7
    References 20

Chapter 3. Spontaneously Occurring Atherosclerosis in Primates 24
    References 37

Chapter 4. Experimental Atherosclerosis 41
    Introduction 41
    Dietary Experiments 43
    Cholesterol Metabolism 55
    Haemodynamic Studies 64
    Hypoxic Injury 65
    The Role of Dietary Carbohydrate 67
    The Effects of Vitamins and other Substances 70
    Various Sorts of Vascular Injury 73
    The Effects of Hormones and Stress 76
    Inhibition of Experimental Atherosclerosis 79
    Regression of Atherosclerotic Lesions 81
    Conclusion 85
    References 85

    Subject Index 96

Acknowledgements

Atherosclerosis research is team work and day by day the team gets larger incorporating more of the aspects of biology and the natural sciences. I am grateful to my local team of assistants, research students and technicians for their help and advice and to the international teams with whom we have close collaboration. In particular my colleagues in the United States and in western Europe. This cooperation enabled a book of this sort to appear.

Most of our work with primates has been with my colleague Dr. A. N. HOWARD and latterly with Dr. D. E. BOWYER. Without their support and skills much of the work would not have been possible.

My secretary, Mrs. TINA GILLINGHAM, has shown astonishing patience and skill and my publishers are to be congratulated on their forbearance on the final production of this work.

G. A. GRESHAM

Preface

Over the past 5 years or so a vast amount of work has appeared about non-human primates in relation to atherogenesis. It is peculiarly opportune to review the situation at the present time particularly because we have arrived at a critical point in the future use of these animals in research. This is because a number of restraints has been imposed by various governments on the sale and exportation of non-human primates for research. These apply especially to the rhesus monkey and the baboon. There is a real need for a summary of the results of experimental work to date and for a detailed appraisal of the experimental work that might be contemplated for the future.

Apart from a few fortunate situations as occur in Malaysia, monkeys are not only scarce but are also expensive to buy, to feed and to breed. The cost of breeding a rhesus monkey, in this country, is now of the order of £ 650. It is not therefore surprising that the majority of animals that are used for research are still imported from the wild state.

The solutions to the problem are not immediately obvious. Already the establishment of large primate centers, as exist in the USA, is one way to tackle the difficulty. Another approach is to ensure the maximum use of material by a number of research groups particularly if the animals are to be used for acute, short-term experiments.

It would also be desirable if most of the experimental work could be devised so as to last for a number of years. However, before embarking on such a programme it is essential to ensure a homogeneous population of animals in order to eliminate the common problem with most experimental animals of interspecies variation. Such long-term experiments would often require a large number of experimental subjects and controls if the results are to be suceptible to analysis at the end of the work.

Perhaps the most difficult problem of all is to decide whether a monkey is essential for the proposed work or whether some more convenient animal such as the rat would be equally satisfactory. In the field of atherosclerosis research many of us have oscillated, over the years, between the use of animals that develop atherosclerosis without provocation, the monkey being such an example, or the rat that is resistant to the production of atherosclerosis.

An important aim of this monograph is to try to resolve this problem of the selection of a suitable experimental animal for research into atherogenesis.

Chapter 1

Atherosclerosis in Man

Occlusive atherosclerosis and thrombosis taken together are vascular disorders that are largely confined to the human primate and it may be argued that 'the proper study of mankind is man' in this field of research.

Atherosclerotic lesions occur in various parts of the vascular tree of non-human primates but are most frequently found in the aorta [1]. Coronary artery disease is much less frequent and is usually of slight degree [2]. However, the sparse occurrence of atherosclerosis in certain human races such as the Masai [3] may sustain the view that all primates are essentially similar so far as the occurrence of atherosclerosis is concerned but that other factors aggravate the disease in certain races of man leading to the production of occlusive and thrombotic disease. Of all creatures birds are most like man in having spontaneous occlusive atherosclerosis [4]. Nevertheless, a variety of good reasons will be adduced in the chapter on experimental atherosclerosis to support the use of non-human primates in this field of research. The intention in this chapter is to give a brief survey of human lesions so that they may be compared and contrasted with those that occur or are experimentally produced in non-human primates.

One of the principal problems in research into atherosclerosis is that of the nature and fate of the fatty streak. These are flat lesions occurring either as streaks or spots and are found most often in the posterior thoracic aorta of persons in their teens, at the origin of the internal carotids and above the aortic valves in the ascending aorta (fig. 1). The difficulty about the fatty streaks is that they do not always correspond in position to the more advanced lesions of atherosclerosis that are found in older people and for this reason some do not regard them as the precursors of atherosclerotic lesions found later in life [5]. This is quite true so far as the aorta is concerned for fatty streaks are most often found in the posterior thoracic part of the vessel whereas advanced atherosclerosis is more frequent in the abdominal section. It is not, however, true of the streaks in the internal carotids where all gradations from fatty streaks to complicated lesions may be found at different ages.

Fig. 1. A view of an opened human thoracic aorta showing extensive fatty streaking of the posterior walls of the vessel.

The fatty streak is a reversible lesion in the experimental animal [6] and it would not, therefore, be surprising if lesions came and went in developing human animals as pressures and circulatory volumes change with increasing age. The view that the fatty streak is not the precursor of the more complex forms of atherosclerosis is therefore not proven. This is an important matter because most of the lesions that occur spontaneously or are induced experimentally in non-human primates are of the fatty streak variety.

In more recent times interest has been revived in the so-called gelatinous lesion that seems to be an intermediate stage between the fatty streak and the later fibrous plaque [7]. This is a grey, translucent, elevation with a faint yellow tint due to the presence of some lipid in it. It is rarely larger than

1 cm across and is distributed like fatty streaks in the thoracic aorta but is also found in the more proximal abdominal aorta as well. It is probably a variant of the lesion which was called intimal oedema by German authors in the early part of the century. The gelatinous lesion can be regarded as either a progressive or regressive stage of the fatty streak, and could account for the discrepancies in the distribution of fatty streaks and more advanced atherosclerotic lesions.

The human fatty streak or spot is an intimal lesion [8]. Several components have been shown in it by various methods of microscopy using visible light, electron microscopy and histochemistry. Lipid is one of the frequent substances and this is usually cholesteryl oleate [9, 10]. The lipid may be intracellular and extracellular in the intima; only later in the disease is it found in substantial quantities in the adjacent media. In more advanced fibrous plaques the nature of the lipid changes: sphingomyelin being the lipid that is most often associated with the sclerotic fibrous plaque [11].

Other substances such as acid mucopolysaccharides collect in the fatty streak [12, 13] and in addition fibrin as well as fibrinogen have been shown to be present [14]. Various authors have considered the aetiological role of these substances some laying more emphasis on lipid accumulation others supporting the view that polysaccharides and even calcium are primarily deposited and so on. There is no firm consensus of opinion, to date, that supports one view or another.

Other constituents of the fatty streak include collagen fibres, elastic fibres and smooth muscle cells [15]. The smooth muscle cell has been demonstrated by electron microscopy [16] and by the use of fluorescent methods it can be shown to contain actomyosin [17]. The features have been shown in atherosclerotic lesions in a number of species as well as in man. This cell is important in the processes of elastogenesis and fibrogenesis. Using immunohistochemical methods serum lipoproteins have been found intracellularly in smooth muscle cells so that they may be also concerned with the deposition of lipids in atherosclerotic lesions [18]. The smooth muscle cell may arise *de novo* or may migrate from the media into the lesion. The significance of smooth muscle cells in atherosclerotic lesions is still by no means clear, though most workers would now agree that they are concerned in the synthesis of scleroproteins [19]. It is often the case in this field of research that electron microscopy throws up more problems than it solves.

The later stages of atherosclerosis are called fibrous plaques where collagen is the preponderating feature and atheroma proper where the bulk of the occluding lesion is formed of lipid. Breaks in the surface of such lesions may occur by intramural haemorrhage or by loss of the endothelial cells that form the internal covering. Any break in the smooth lining of the artery may predispose to thrombosis either by the production of a physically rough surface or by the exposure of subendothelial collagen to both of which platelets will adhere readily [20]. In this way occlusive thrombosis may be produced. Medial degeneration with loss of elastic tissue and muscle may be the result of ischaemia brought about by intimal thickening for there is evidence that the inner media in man depends upon luminal diffusion of nutrients for its supply [21]. These more complex forms of atherosclerosis tend to be found frequently in man and only very rarely in other primates. It is most unlikely that further knowledge about the pathogenesis of atherosclerosis will come from a study of advanced lesions. Many processes have been involved in their formation; the important question is to discover how they began. For an answer to this problem the earliest stage must be studied and it is most likely that this is the fatty streak or spot, or the gelatinous lesion.

Despite the great bulk of research in this field the prevalent theories of atherogenesis are very similar to those that existed in the last century. There are those who support the thrombogenic hypothesis first produced by VON ROKITANSKY [22]. The notion is that formed elements of the blood are deposited in the intima and become incorporated into it to form the plaque. These elements may be fibrin or platelets or thrombus. It is, however, most unusual for this to happen without prior endothelial injury though some authors have shown fibrin deposition on normal vessels [23]. The thrombotic tendency in primates other than man is not great; thrombi are rarely found in primate arteries unless there is traumatic or parasitic damage so that it seems unlikely that the thrombogenic process is an important factor in non-human primate atherosclerosis.

The second principal hypothesis is concerned with the imbibition of lipid into the intima. Some lipids are histotoxic and cause fibrosis [24] but most people believe, as did VIRCHOW who originally propounded the notion, that some degree of endothelial damage was needed before lipid would collect in the intima. The nature of the injury has not yet been specified. The final hypothesis that is really no more precise than the other two, may be called the single or multiple insult hypothesis. It emphasises the idea that a variety of injury to arteries induced by chemical, traumatic or other means may result in atherosclerosis. Further accumulations of lipid, polysaccharide, fibrinogen and so on are all secondary to this initial insult. The atherosclerotic plaque might be regarded as a scarring process that follows intimal injury and subsequent inflammation. As with scars in other parts of an animal the process may be followed by lipid accumulation and calcification.

In subsequent chapters of this book we shall examine the evidence for the various theories of atherogenesis as obtained from studies of spontaneously occurring and experimentally induced atherosclerosis in non-human primates.

References

1 GRESHAM, G.A. and HOWARD, A.N.: Vascular lesions in primates. Ann. N.Y. Acad. Sci. 127: 694–701 (1965).
2 STRONG, J.P.; EGGEN, D.A.; NEWMAN, W.P., and MARTINEZ, R.D.: Naturally occurring and experimental atherosclerosis in primates. Ann. N.Y. Acad. Sci. 149: 882–894 (1968).
3 HO, K.J.; BISS, K.O.; BELMA, M.; LEWIS, L.A., and TAYLOR, C.B.: The Masai of East Africa. Some unique biological characteristics. Archs Path. 91: 387–410 (1971).
4 PRITCHARD, R.W.: Spontaneous atherosclerosis in pigeons; in ROBERTS and STRAUS Comparative atherosclerosis, pp. 45–50 (Harper & Row, New York 1965).
5 MITCHELL, J.R.A.; SCHWARTZ, C.J., and ZINGER, A.: Relationship between aortic plaques and age, sex and blood pressure. Br. med. J. i: 205–209 (1964).
6 RODBARD, S.; PICK, R., and KATZ, L.N.: The rate of regression of hypercholesteremia and atherosclerosis in chicks. Effect of diet, pancreatectomy, estrogens and thyroid. Circulation 10: 597 (1954).
7 SMITH, E.B. and SLATER, R.S.: Lipids and low density lipoproteins in intima in relation to its morphological characteristics. Ciba Fdn Symp., vol. 12, pp. 39–62 (Van Gorcum, Assen 1973).
8 STRONG, J.P. and MCGILL, H.C., jr.: The natural history of coronary atherosclerosis. Am. J. Path 40: 37–49 (1962).
9 WERTHESSEN, N.T.; NELSON, W.R.; JAMES, A.T., and HOLMAN, R.L.: Composition of fatty acids in cholesterol esters derived from normal and abnormal intima. Circulation 20: 972 (1959).
10 WAHLQVIST, M.L.; DAY, A.J., and TUME, R.K.: Incorporation of oleic acid into lipid by foam cells in human atherosclerotic lesions. Circulation Res. 24: 123–130 (1969).
11 SMITH, E.B.; EVANS, P.H., and DOWNHAM, M.D.: Lipid in the aortic intima. The correlation of morphological and chemical characteristics. J. Atheroscler. Res. 7: 171–186 (1967).
12 GRESHAM G.A.; HOWARD, A.N., and KING, A.J.: A comparative histopathological study of the early atherosclerotic lesion. Br. J. exp. Path. 43: 21–23 (1962).
13 KLYNSTRA, F.B.; BÖTTCHER, C.J.F.; MELSEN, J.A. VAN, and LAAN, J. VAN DER: Distribution and composition of acid mucopolysaccharides in normal and atherosclerotic human aortas. J. Atheroscler. Res. 7: 301–309 (1967).
14 WOOLF, N. and CRAWFORD, T.: Fatty streaks in the aortic intima studied by immunohistochemical technique. J. Path. Bact. 80: 405–408 (1968).

15 WISSLER, R.W.: Arterial medial cell. Smooth muscle or multifunctional mesenchyme. J. Atheroscler. Res. *8:* 201–213 (1968).
16 HAUST, M.D. and MORE, R.H.: Mechanism of fibrosis in white atherosclerotic plaque of human aorta. An electron microscopic study. Circulation *34:* 14 (1966).
17 KNIERIM, H.J.: Immunohistochemical studies on the significance of smooth muscle cells in the pathohistogenesis of human arteriosclerosis. Beitr. Pathol. *141:* 4–18 (1970).
18 KAO, V.C. and WISSLER, R.: A study of the immunohistochemical localisation of serum lipoprotein and other plasma protein in human atherosclerotic lesions. Expl. molec. Path. *4:* 465–479 (1965).
19 POOLE, J.C.F.; CROMWELL, S.B., and BENDITT, E.P.: Behaviour of smooth muscle cells and formation of extracellular structures in the reaction of arterial walls to injury. Am. J. Path. *62:* 391–413 (1971).
20 MUSTARD, J.F.; MURPHY, E.A.; ROWSELL, H.C., and DOWNIE, H.G.: Platelets and atherosclerosis. J. Atheroscler. Res. *4:* 1–28 (1964).
21 FRENCH, J.E.: The structure of arteries; in WOLF The artery and the process of atherosclerosis, pp. 1–52 (Plenum Press, New York 1971).
22 ROKITANSKY, C.V.: A manual of pathological anatomy, vol. 3, p. 261 (Sydenham Society, London 1852).
23 MORE, R.H.; MOVAT, H.Z., and HAUST, M.D.: Role of mural fibrin thrombi of the aorta in genesis of arteriosclerotic plaques. Archs Path. *63:* 612–620 (1957).
24 HARLAND, W.A.; SMITH, A.G., and GILBERT, J.D.: Tissue reaction to atheroma lipids. J. Path. *111:* 247–253 (1973).

Chapter 2

Structure and Function of the Normal Vasculature

As is so often the case in many aspects of experimental pathology studies of disease precede observations on the normal organs or tissues that are affected. A great deal of work has been done on atherosclerosis in primates but the literature on normal vessels, heart and vascular function is, by comparison, sparse. This chapter reviews some of the studies of the normal vasculature of primates and of other animals for it is in this field of work that some of the clues to an understanding of abnormality can be sought and obtained. WISSLER [1] emphasised this matter in 1967 but this salutary warning has not yet generated a great deal of work on primate vascular physiology.

Much work on normal structure still remains to be done; it is perhaps significant that ROHLES' [2] biography of the chimpanzee has as many entries on psychobiology as of anatomy. Once again this reflects the tendency in recent times to study the abnormal rather than the normal.

Many of the clues to the susceptibility of different vessels and parts of vessels to disease lie in a study of vascular structure and also in an understanding of the changes in blood pressure and in blood composition within the vessel itself. For example, a peculiar feature of the human aorta, which is especially susceptible to atherosclerosis, is the discrepancy in the number and thickness of the lamellar units as between thoracic and abdominal parts [3]. This difference does not exist in prenatal life but gradually develops throughout the postnatal period. The increase in thickness of the thoracic aorta is due to the increase in number of lamellar units; each unit consisting of smooth muscle cells attached to coarse elastic fibres supported by collagen and set in a matrix of mucopolysaccharide containing delicate fibres of elastin. Increased thickness of the abdominal aorta with age is due to thickening of lamellar units rather than an increase of number. This creates a disparity in tension within each unit in the thoracic and abdominal aorta. The tension in the abdominal units being greater than that in the thoracic aorta.

It may well be that these differences in unit tension will go part of the way to explain the greater susceptibility of the primate abdominal aorta, and in particular that of man, to develop atherosclerosis.

Differences in structure between thoracic and abdominal aortic segments are not clearly evident at birth but with advancing age there is a progressive increase in medial thickness and a loss of distensibility. In addition to the medial changes there is also fibromuscular intimal thickening, fragmentation of medial elastic tissue and an increase of stainable collagen in the media. As we have already indicated the thickness of the media of both thoracic and abdominal parts of the aorta doubles from the newborn period to the adult stages: in the thoracic aorta this is largely due to a multiplication of the medial elastic lamellae whilst in the abdominal aorta the principal cause of medial thickening is widening of the existing medial elastic lamellae [4]. Not only is the stress per unit of lamella greater in the abdominal than in the thoracic aorta, but the blood supply to the larger abdominal units is no greater than that in the thoracic aorta.

The distribution of the vasa vasorum in the different parts of the aorta in relationship to the number and thickness of elastic lamellae is clearly important because the combination of increased stress per unit coupled with a relative diminution of blood supply can be important in the genesis of disease.

The inner zone of the thoracic and abdominal aortic media of man and other mammals is an avascular territory of about 28 lamellar units in thickness [5]. The outer media is nourished by vasa vasorum in other mammals but in man the abdominal aortic media is totally avascular. This fact may account for enlargement of existing lamellar units, rather than the formation of new ones, as the media thickens with age.

The cause of medial thickening is in part due to rising blood pressure during life. This follows a pattern which is characteristic for each particular species of animal [6] and there seems to be good reason to associate these medial changes with the usual susceptibility of the abdominal part of the primate aorta to atherosclerosis.

The thickness of the aortic media obviously varies with the size of the animal. For animals less than 5.6 kg in weight the maximum number of lamellar units is 24. So that in the rhesus monkey the number is 23.5 whereas in the mouse the number of units is 5. Up to 24 U there are no medial vasa vasorum in the vessel wall; after that figure is reached vasa can be found in the outer media.

The number of lamellar units in the aortic wall of mammals increases with age. In all of them there is an inner avascular zone which may be up to 29 U in thickness and it seems that this is the upper limit of vessel wall thickness that can be nourished by diffusion from the lumen.

In animals with few lamellar units the process of stripping the adventitia has no effect upon the diffusion of nutrients through the vessel wall. For example there is no effect upon the diffusion of cholesterol through the rat or rabbit aorta after adventitial stripping [7]. In large animals, however, occlusion of the vasa vasorum results in focal necrosis of the middle media [8]. Because the width of the inner vulnerable zone increases with the size and weight of the animal it follows that large primates are more likely to resemble man at least in aortic behaviour than are the smaller primates. A wide variety of differences in the aortic lamellar unit structure is seen throughout the animal kingdom [9]. These variation can frequently be related to the tendencies to develop atherosclerotic disease.

Not only does the thickness of the arterial wall change with age but there is evidence that changes also occur in the nature of the elastic components that are so frequently involved when atherosclerosis develops. Whether these changes that occur with age are pathological or not is a matter for debate at the present time. In the embryo elastic fibres appear to be derived from smooth muscle cells or fibroblasts. They synthesise two components: one is a proelastin or tropoelastin [10] and the other is a structural glycoprotein [11]. Electron microscopy shows the former as amorphous droplets and the latter as classical microfibrils [12]. Together they assemble to form the primitive elastic fibres. The glycoprotein component disappears with increasing age and the mature elastic fibril remains as the original units of tropoelastin cross-linked by desmosine and other amino acids [13]. The more elaborate tertiary and quaternary arrangements of elastic fibres seem to be caused by hydrophobic interactions. This latter point is of considerable importance because of the donating effect of lipids on the process of hydrophobic stabilisation. It may be that this explains one of the mechanisms whereby lipids may damage the arterial wall particularly if inadequate perfusion of the inner vascular media occurs as in larger animals.

Age changes in elastic tissue have been summarised recently [14] and it seems that a number of processes are involved. Possible important mechanisms here may be the formation of anti-elastin antibodies derived from emergent forbidden clones of antibody-forming cells which may arise for one reason or another. These antibodies by fixing complement may lead to the liberation of complement-dependent degrading enzymes from lysosomes which destroy the elastic tissue.

Another factor damaging elastic tissue may be through antigen-antibody complexes causing platelet clumping with the subsequent liberation of platelet proteases that digest elastin. These processes may be an inevitable accompaniment of aging in primates or they may be directly caused and may hence be controllable. Whichever be the case it is clear that further studies of the biochemistry of vascular elastin is needed to elucidate the mechanisms of atherogenesis.

Elastolytic enzymes have been known for many years and have been found in the pancreas and isolated from the blood [15]. What part, if any, they play in the processes of growth and remodelling of elastic vessels still remains to be determined and the question whether they are in any way involved in atherogenesis is still unsolved [16]. There are small amounts of elastoproteinase and elastomucase in extracts of aorta [15]. The former acts on the protein constituents and the latter on the mucopolysaccharides of elastin. Evidence for the presence of the lipopolytic enzyme elastolipoproteinase in the vessel wall has not yet been forthcoming. At least one inhibitor of elastoproteinase has been found in this situation suggesting that a balanced mechanism exists in the vessel wall which may control the quantity of elastic tissue which is present. The existence of monomeric and dimeric forms of elastoproteinase further complicates the issue in that the former acts upon elastin molecules cross-linked by calcium whereas the latter attacks the free carboxyl groups of elastin. The monomeric forms seems to act preferentially on adult elastin so that defects in elastogenesis or elastic tissue remodelling may result not only from disturbances in the quantity of enzymes but also in the qualitative differences they may exist.

In certain situations as for example the uterine arteries [17] it appears that successive waves of elastic fibre deposition tend to occur with repeated pregnancies and that the ability to remodel or remove these layers becomes increasingly difficult as the animal gets older. Once again this reflects the varying response of the vessel wall in different situations in the body. GILLMAN and HATHORN [18] regard these waves of 'elastosis' as part of a repair reaction in the vessel wall and state 'degeneratory vascular diseases will probably be effectively understood only if and when knowledge concerning repair processes in connective tissues generally is applied to the analysis of vascular injury in particular'. This observation, made in 1959, is still as true, even more so today. They studied the effects of feeding *Lathyrus odoratus* on the regeneration of elastic tissue in the damaged artery and emphasise the fact that extractable hexosamines increase in amount before the deposition of reticulin or collagen occurs. This is a process which is similar to that seen in healing wounds and one which is common to most tissues following damage. They also discuss the mechanisms of remodelling of elastic tissue in the arterial wall as the vessel grows and suggest that the dissecting aneurysms that are found in lathyrism are the result of lysis of the inner elastic membrane which occur as a normal event but with Lathyrus intoxication is not accompanied by further growth of new elastic tissue: this latter process is inhibited by β-amino propionitrile.

Young animals are much more susceptible to Lathyrus extracts than older ones and this indicates the need to study young developing vessels when seeking the causes of atherosclerosis. In man the maximum vascular remodelling occurs between 10 and 20 years of age and it may well be that the substrate for atherosclerosis is laid at this period of life. Despite objections that have been made in times past it seems clear that the use of young growing primates is logical if the early stages of atherogenesis are to be elucidated and prevented.

Changes in elastic tissues such as duplication and fragmentation of these membranes are a constant feature in most atherosclerotic lesions that are found in primates. But the problem still remains whether or not the changes are primary or secondary. Calcification of the vessel wall may be caused by degeneration of elastin or by accumulations of collagen or mucopolysaccharides [19]. Again there are different views about this. Some support the view that calcification is secondary, others that it is the primary event starting in mitochondrial membranes and leading to cellular damage. Not all degenerate elastin calcifies that in the skin is a good example. By a similar analogy bone is not always found in places where there is calcium and collagen and other constituents for its formation. Elastic tissue cannot be regarded as a homogenous entity throughout the animal body and it is clear that research into atherogenesis should include the special properties of arterial elastica which causes it to differ from elastic tissue elsewhere in the body.

The effects of vascular mucopolysaccharides have also been the subject of much controversy. Most of these substances can be depolymerised enzymatically with testicular hyaluronidase and in moderate amounts they are a feature of the normal vessel being more conspicuous in the young developing vessel than in that of the normal adult. This extracellular collection of mucopolysaccharides is an important feature that distinguishes the smooth muscle of the arterial wall from all other smooth muscle-containing tissues [20]. It may be that the principle function of the material is to bind extracellular sodium ions in the arterial wall which maintains the reactivity of the vessel [21]. The presence of mucopolysaccharide in excess occurs in hypertensive vascular disease [22] and of course atherosclerosis is more often found when there is sustained hypertension. The difficult question is whether the mucopolysaccharide is the substrate which leads to further vascular damage and to atherosclerosis ultimately or whether its presence is merely a reflection of the increased ion-binding capacity of the vessel wall that might maintain increased vascular reactivity in hypertensive disease. Studies of the aorta in coarctation [23] suggests that the mucopolysaccharide collects because of increased arterial wall metabolism brought about by increased intraluminal pressure. Somewhat in favour of this view is the rise in mucopolysaccharide content that follows large intravenous doses of epinephrine given twice daily to rabbits [24]. If the vessels are treated with epinephrine *in vitro* then the mucopolysaccharide content falls [25].

Reactivity and active metabolism are features of vessels that are especially liable to develop atherosclerosis. Thus, the proximal aorta of the baboon is largely elastic in nature in order to act as a systemic pressure reservoir [26]. The abundant elastic tissue in this part of the aorta resists vascular deformation and transmits the pressure effect as a wave along the vessel to be taken up by the more muscular distal aorta and its branches. It is the reactive contractile distal aorta and its more muscular off-shoots that are peculiarly liable to develop atherosclerotic disease. Similar differences in structure of the aorta have been noted in rhesus monkeys [27], primitive prosimian primates [28] and in many other mammals and birds [9].

Another factor that needs to be considered when susceptibility to atherosclerosis is being studied is the ratio of collagen to elastin in the normal vessel wall. A high proportion of collagen reduces distensibility and consequently augments the effects of any occlusive intimal lesion that might be present. The high ratio of collagen to elastin in coronary arteries is an important consideration in this respect not only because it makes the vessel wall more rigid but it also affects the ability of coronary arteries and coronary collateral vessels to dilate in order to compensate for occlusions that may develop elsewhere in the system [29].

More study is also needed of the ways in which collateral circulations might develop especially in the coronary circulation after occlusion of one of the main vessels have occurred. Recent studies on the partially occluded retinal circulation in man, pig and the monkey has demonstrated the ease with which new collaterals develop from primitive thinwalled vascular channels. These vessels readily develop new muscle in their walls so that it seems that the anatomical basis for new collateral vessels is not a problem [30]. However, there is evidence that collateral circulation is not only dependent upon mechanical control but that it may be influenced by hormonal factors as well [31]. The more rapid development of pulmonary collateral vessels in puppies as compared to adult dogs supports this view [32]. Experiments with rats in which one branch of the pulmonary artery was ligated indicated that growth hormone and cortisone had effects upon the development of pulmonary collaterals that were independent of any changes in the weight of the animals. Cortisone tended to reduce the appearance of collaterals whereas growth hormone tended to promote their development [31].

Considerable controversy still exists about the dynamics of flow in the coronary collateral circulation after coronary artery occlusion and myocardial infarction. Work in the dog using heated probes and krypton and xenon clearance methods have shown a decline of flow in the infarcted area with little change of flow or pressure in the adjacent dilated collaterals. Similar experiments in patas and African Green monkeys and in the baboon has shown substantially the same results. The collateral hyperaemia near the infarcted area is potentiated by blockade of two receptors [33].

Different morphological patterns in the distribution of coronary arteries exist in mammals and the ideal animal for research into the factors affecting coronary artery disease should have a similar coronary artery anatomy to that found in man. Dogs differ from man, monkey and pig by the presence of numerous anastomotic channels between the main coronary arteries.

By means of vinyl casts and other methods a variety of studies have been made on the distribution of coronary arteries in man and in the Old and New World monkeys [34]. The numbers of animals used is often small and the results may therefore not be valid for all members of the species examined. In the human heart the right coronary artery preponderates in about 80% of subjects examined. This method of evaluation determines whether the right or left artery supplies the posterior ventricular sulcus. Rhesus monkeys [35] and chimpanzees [36] fairly consistently show a right coronary artery preponderance. The wooly and squirrel monkeys have preponderant right coronary arteries as well but the vessels are shorter and extend about half way from the base of the heart towards the apex in the posterior sulcus. Howler monkeys, spider monkeys, orang-utan and gorilla have left coronary preponderance whereas the Kenya baboon is unique in having both right and left arteries contributing equally to the supply of the posterior sulcus [37].

Conflicting statements about coronary artery distribution do, however, exist and may be the result of examining too small a sample of animals. KANTARIA [38] investigated 382 hearts from macaques, baboons, patas and vervet monkeys by radiographic perfusion methods and showed that much greater variations in coronary artery distribution occurred than are seen in man.

Attempts to use electrocardiographic methods for the study of induced ischaemic myocardial disease in primates must take these different arterial distributions into account. The correlation between ischaemic changes as seen in the electrocardiogram and obstruction of a particular artery will not be possible unless such variations are carefully considered. This field of research is made even more difficult by the debate that continues over the pathological basis for many electrocardiographic abnormalities. For example some claim to be able to demonstrate lesions of the left bundle branch in cases of left bundle branch block others do not [39]. UNGER *et al.* [40] illustrate lesions of the left bundle branch in such cases but the problem in man and for that matter in some of the other primates is the difficulty in demonstrating the conducting system clearly and also in sorting out pathological fibrosis from the dense collagen that occurs normally at the base of the heart.

There are also electrocardiographic variations from one species to another and this is true amongst the primates as well as for other animals. As a general rule it is true that the smaller the animal and the faster the cardiac rate the more likely there is to be an elevation of the ST segment. This is a normal event and must not be construed as indicating ischaemia [41].

Considerable variation in the P wave has been demonstrated in normal squirrel monkeys. P waves are often tall and spiked in these animals and similar changes have been recorded in rhesus monkeys, Celebes apes and Japanese macaques [42]. P waves of squirrel monkeys were also shown to vary in form and amplitude and were occasionally inverted. This occurred in the anaesthetised animal and the effects of anaesthesia cannot be excluded in this animal. This is likely to be the case because the changes were transient and were not found in subsequent records taken a year later.

Studies of other primates have shown that considerable variations exist in the electrocardiogram. The phenomenon of wandering pacemaker has been observed in a conscious chimpanzee [43] and prolongation of the QRS has been described in rhesus and cebus monkeys as well as in the squirrel monkey. Variations in the T wave indicating differences of repolarisation are seen in a wide range of non-human primates; these should be considered to be a normal phenomenon because they have not been associated with demonstrable lesions in the heart [44, 45]. However, the basic arrangement of the conducting system seems to be similar in man and other primates more so than that in other animals such as dog and calf. Experiments have been done in baboons where incisions were made into the left side of the ventricular septum and this resulted in electrocardiographic changes similar to those seen in septal ischaemic damage in man [46]. Provided variations in the electrocardiogram are anticipated between one primate and another electrocardiograms can be a useful object in the study of occlusive vascular disease in the primate.

Slight quantitative differences in the cardiograms of macaques were observed when made under light anaesthesia with Nembutal but the main effects of the barbiturate were occasional ectopic beats, slowing of cardiac rate and proportional prolongation of various electrocardiographic intervals. Similar changes have been observed in baboons anaesthetised with other barbiturates [47] and in animals tranquillised by Sernyl [48]. The changes consisted of elevation of the ST segment and inversion of the T wave in leads II, III and AVF [49]. Other parameters were measured during these studies including phonocardiograms, apexcardiograms, aortic, pulmonary and intracardiac pressures. The salient features of both phonocardiograms and apexcardiograms were nearly identical with human records. The general wave forms and absolute magnitudes of blood pressure were also similar to those in normal man. Another report [71] by the same authors showed that aortic blood flow and the cardiovascular responses to spontaneous activity and exercise were essentially the same as those in normal human beings.

Of all non-human primates the largest number of physiological studies have been done with *Macaca mulatta*. The animal is of convenient size and is easy to use for such measurements. A large number of data have been accumulated at the Oregon Primate Centre [51]. These show that metabolism, oxygen uptake, respiration rate, minute volume, lung volume, heart rate, lung and blood gas tensions and so on are essentially proportionally similar to those of man.

Because of the fact that many believe atherosclerosis and its complications to be caused by insudation of plasma constituents into the vessel wall considerable interest has centred on the comparisons of metabolism and chemistry of blood constituents in the various primate species.

Early studies were concerned with haematological estimations primarily because of an interest in the effects of radiation exposure [52]. Difficulties arose originally because of the small numbers of animals that were used but since that time much more work has been done in various sorts of macaques and baboons [53]. Newly imported macaques were found to have a moderate anaemia and leucocytosis with an elevation of the plasma globulin levels. These changes were associated with respiratory and enteric infections which required some three courses of therapy before the values became normal [54]. Similar changes have been observed in baboons [55]. Relatively little has been written about the cellular constituents of the blood in primates. More will be said subsequently about platelet behaviour in these animals.

Biochemical studies have been more extensive and include reports upon levels of lipids, electrolytes, enzymes and so on [56]. Cholesterol and blood lipid levels tend to be lower than those of man. In the baboon, for example, the cholesterol level is round about 100 mg/100 ml and similar levels were recorded for macaques [57]. Sequential measurements of lipid levels in control animals in any experiment are essential particularly when lipids are one of the variables being studied. The reason for this is that a small but significant rise in cholesterol level occurs when the animal is taken into captivity. The effect of restraint has been studied on the serum cholesterol levels of male and female rhesus monkeys. This showed a close relationship between levels of the sterol and the period of restraint. Similar alterations have been shown for squirrel monkeys kept in the laboratory as compared to newly trapped animals. The cholesterol levels rose after a period of captivity but tended to fall again as the animals became acclimatised to the environment [58]. These changes were entirely accounted for by a rise in β-lipoprotein cholesterol; there was no change in the cholesterol transported by the α-lipoproteins. This altered ratio of β- and α-lipoproteins approximates towards the condition that exists in man. The precise way in which captivity effects this change is not clear but further studies may illuminate the relationship of 'stress' of captivity and atherogenesis; at present a much debated topic.

As well as serum lipids serum enzymes are important indicators of the presence of occlusive vascular disease and myocardial necrosis in the experimental animal. Considerable variation in serum enzyme levels within 'normal' limits was found in *M. mulatta*. This may in part, be another effect of stress in the environment. For example, studies on SGO-T in rhesus monkeys show such a relationship [59]. Glutaric acid is present in most tissues and in serum in equilibrium with its keto acid α-ketoglutaric acid; this is a substrate in glycolysis and increases in amount with exercise or muscular stress. Exercise leads to a rise in SGO-T levels which then return to normal though exercise continues. More stressfull situations including injections of epinephrine led to a sustained increase in SGO-T levels. It was noteworthy that the response varied from animal to animal, some being more susceptible than others. Similar variations in response to all sorts of stimuli are seen within species of monkeys. The phenomenon has been described after cholesterol feeding in the squirrel monkey; the degree of hypercholesterolaemia that develops varies greatly from one animal to another [60]. The question

Structure and Function of the Normal Vasculature 17

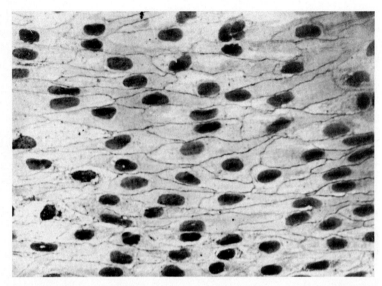

Fig. 2. A 'Häutchen' preparation of aortic endothelium. The intercellular borders are outlined by silver. Methylene blue. × 400.

of 'responders' and 'non-responders' is a vitally important matter in the field of atherosclerosis research. In order to minimise the problem one must have a large number of experimental animals and controls or use highly inbred stock. Much of the work that is done with small groups of primates is likely to be invalidated because of this problem and attention has been directed towards the establishment of large primate centres in order to overcome it. In this regard the United States has led the field by creating such facilities [61].

Finally it is important to consider the reacting surface in atherogenesis: this is the endothelial lining of the vessels. Very little work has been done on this matter in primates. Endothelial cells can be studied as a static picture in sections but this is unsatisfactory as the cells are so thin and so few can be seen this way. Another approach is to strip the endothelial surface of the inner lining of the artery; these are so-called 'Häutchen' preparations. In this way the endothelial cells and their junctions can be viewed 'en face' (fig. 2). A variety of methods are available for producing this endothelial sheet; the most satisfactory uses cellulose acetate paper [62]. Another way of looking at endothelium at a higher power is by the methods of scanning electron microscopy. Some authors have described gullies and folds in the

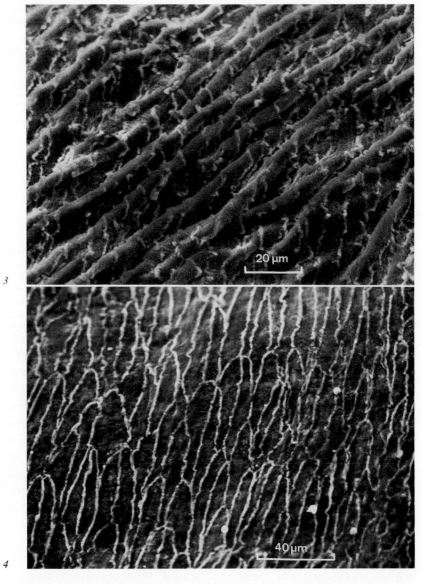

Fig. 3. Scanning electron micrograph of aortic endothelium. This preparation was not fixed under pressure. Note the longitudinal ridges and gullies traversed by cement lines. Courtesy of Dr. PETER DAVIES.

Fig. 4. A similar preparation to figure 3, but this time the sample was fixed under pressure. The gullies and ridges seen in figure 3 have disappeared. Courtesy of Dr. PETER DAVIES.

surface with the nuclei buried deeply in the gullies [63]. Only when the cells round up to do the gullies vanish and the endothelial cells pull apart. The functions of these folds in the endothelial surface at present remain obscure. They have been demonstrated in man, monkeys, rabbits, dogs and guinea pigs in the large arteries and veins. Obliteration of the folds followed the infusion of substances such as cholesterol, epinephrine, norepinephrine bradykinin and angiotensin. It seems that the elevation of nuclei out of the depths of the folds exposed them to injury and cholesterol administration aggravated this by producing fatty streaks or spots on the endothelial surface. More recent work by BOWYER and his group [64] has shown that the gullies described by many authors are most likely an artefact of fixation. If the vessels are fixed at mean arterial pressure and the endothelial surface subsequently studied by scanning electron microscopy these gullies are no longer visible. The effects of perfusion with various substances that have already been discussed may therefore be entirely non-specific and created by the pressure of perfusion (fig. 3, 4).

A more dynamic view of the endothelial surface can be achieved by subjecting the cells to the effects of tritiated thymidine and then, by autoradiographic methods, studying the number of cells in incipient mitosis. This enables a look at the ways in which endothelial cells respond to injury [65] and has been done in monkeys. Work in other animals has shown that uptake of labelled thymidine is more rapid at points of stress in the arterial tree, namely at branches, trifurcations and bifurcations where atherosclerotic lesions are known to occur [66].

Similar studies have been made with convential electron microscopy showing that certain stimuli such as low or high pH, anoxia and solutions of low osmolality led to an opening of the intercellular junctions [67–69]. It is, of course, unlikely that such conditions might prevail in the intact artery *in vivo,* nevertheless, studies of this sort are an important prelude to the discovery of less vigorous but more significant factors that might predispose to the accumulations of blood products in the arterial intima of man.

Apart from recent developments and applications of electron microscopy, histochemistry at the light and electron microscope levels and such techniques as micro-probe analysis, research into atherogenesis has gradually turned from the morphologic to the more functional aspects of disordered arteries.

Recent interest in flow and shear as factors in lipid deposition in the vessel wall have opened new avenues for research [70]. This work is aided by the applications of radio-telemetry for the study of flow in the major vessels of unrestrained animals including monkeys [71]. The concept of factors in the blood itself has been enlarged to include the possibility of immunological injury as a factor in atherosclerosis production [72] and the hypothesis that atherosclerotic lesions are monoclonal in origin [73] supports this view.

Atherosclerosis research is now a compound of many disciplines in biology all of which demand a more fundamental knowledge of normal vascular behaviour if further progress is to be made.

References

1 WISSLER, R.W.: The arterial medial cell, smooth muscle, or multifunctional mesenchyme? Circulation *36:* 1–4 (1967).
2 ROHLES, F.H.: The chimpanzee: a topical bibliography. Technical documentary reports No. A.R.L.–T.D.R.–63–27 (1963) and A.R.L.–T.R.–67–4 (1967).
3 WOLINSKY, H. and GLAGOV, S.: Comparison of abdominal and thoracic aorta medial structure in mammals. Deviation of man from the usual pattern. Circulation Res. *25:* 677–686 (1969).
4 WOLINSKY, H.: Comparison of medial growth of human thoracic and abdominal aortas. Circulation Res. *27:* 531–538 (1970).
5 WOLINSKY, H. and GLAGOV, S.: Nature of species differences in the medial distribution of aortic vasa vasorum in mammals. Circulation Res. *20:* 409–421 (1967).
6 DAWES, G.S.: Foetal and neonatal physiology, p. 98 (Year Book, Chicago 1968).
7 ADAMS, C.W.; BAYLISS, O.B.; DAVISON, A.N., and IBRAHIM, M.Z.M.: Autoradiographic H^3 evidence for the outward transport of H^3 cholesterol through rat and rabbit aortic wall. J. Path. Bact. *87:* 297–304 (1964).
8 WILENS, S.L.; MALCOLM, J.A., and VAZQUEZ, J.M.: Experimental infarction (medial necrosis) of the dog's aorta. Am. J. Path. *47:* 695–711 (1965).
9 BERRY, C.L.; GERMAIN, J., and LOVELL, P.: Comparison of aortic lamellar unit structure in birds and mammals. Atherosclerosis *19:* 47–59 (1974).
10 SANDBERG, L.B.; WEISSMAN, N., and SMITH, D.W.: The purification and partial characterization of a soluble elastin like protein from copper deficient porcine aorta. Biochemistry *8:* 2940–2945 (1969).
11 MOCZAR, M.; MOCZAR, E., and ROBERT, L.: Composition of glycopeptides obtained by proteolytic digestion of the media of porcine aorta. Atherosclerosis *12:* 31–40 (1970).
12 ROSS, R. and BORNSTEIN, P.: The elastic fibre. J. Cell Biol. *40:* 366–381 (1969).
13 ROBERT, L.; ROBERT, B., and ROBERT, A.M.: Molecular biology of elastin as related to aging and atherosclerosis. Expl. Gerontol. *5:* 339–356 (1970).
14 HALL, D.A.: The aging of connective tissue. Introductory lecture. Expl. Gerontol. *3:* 77–89 (1968).
15 LOEVEN, W.A.: Elastolytic enzymes in the vessel wall. J. Atheroscler. Res. *9:* 35–45 (1959).
16 LOEVEN, W.A.: The possible implication of the enzymes of the elastase complex in the process of atherosclerosis. Archs Mal. Cœur *59:* suppl. 3, pp. 96–106 (1966).

17 ALBERT, E.N. and BHUSSRY, B.R.: The effects of multiple pregnancies and age on the elastic tissue of uterine arteries in the guinea pig. Am. J. Anat. *121:* 259–269 (1967).
18 GILLMAN, T. and HATHORN, M.: Post-natal vascular growth and remodelling in the pathogenesis of arterial lesions. Schweiz. Z. Path. Bakt. *22:* 62–87 (1959).
19 YU, S.Y. and BLUMENTHAL, H.T.: The classification of elastic fibre. Lab. Invest. *12:* 1154–1162 (1963).
20 BUNTING, C.H. and BUNTING, H.: Acid mucopolysaccharides of the aorta. Archs Path. *55:* 257–264 (1953).
21 PALATY, V.; GUSTAFSON, B., and FRIEDMAN, S.M.: Sodium binding in the arterial wall. Can. J. Physiol. Pharmacol. *47:* 763–770 (1969).
22 MANLEY, G.: Changes in vascular mucopolysaccharides with age and blood pressure. Br. J. exp. Path. *46:* 125–134 (1965).
23 HOLLANDER, W.; MADOFF, I.M.; KRAMSCH, D., and YAGI, S.: Arterial wall metabolism in experimental hypertension of coarctation of the aorta; in Hypertension, vol. 13, p. 191 (American Heart Ass., New York 1965).
24 MILCH, L.J. and LOXTERMAN, P.B.: Aortal mucopolysaccharide changes after epinephrine administration in rabbits. Proc. Soc. exp. Biol. Med. *116:* 1125–1126 (1964).
25 HOLLANDER, W.; YAGI, S., and KRAMSCH, D.M.: *In vitro* effects of vasopressor agents on the metabolism of the vascular wall; in Hypertension, vol. 12. Circulation *29/30:* suppl. 11, p. 11 (1964).
26 KATZBERG, A.A.: The attenuation of elastic tissue in the aorta of the baboon. *Papio doguera.* Anat. Rec. *154:* 213–220 (1966).
27 LAPIN, D.A. and YAKOVLEVA, L.A.: Comparative pathology in monkeys (Thomas, Springfield 1963).
28 AHMED, M.M.: Microscopic anatomy and the attenuation of elastic tissue in the aortic wall of slow loris *(Nyceticebus coucang coucang).* Folia primatol. *8:* 290–300 (1968).
29 FISCHER, G.M. and LLAURADO, J.G.: Connective tissue composition of human coronary artery and its relationship to divalent cation content. Angiology *22:* 31–36 (1971).
30 DOLLERY, C.T.: The formation of collaterals. Post-grad. med. J. *44:* 28–31 (1968).
31 MEFFERT, W. and LIEBOW, A.A.: Hormonal control of collateral circulation. Circulation Res. *18:* 228–233 (1966).
32 LIEBOW, A.A.; HARRISON, W., and HALES, M.R.: Experimental pulmonic stenosis. Bull. int. Ass. med. Mus. *31:* 1–23 (1950).
33 GRAYSON, J. and IRVINE, M.: Myocardial infarction in the monkey: studies on the collateral circulation after acute coronary occlusion. Cardiovasc. Res. *2:* 170–178 (1968).
34 COLBORN, G.L.: The gross morphology of the coronary arteries of the common squirrel monkey. Anat. Rec. *155:* 353–367 (1966).
35 CHASE, R.E.: The coronary arteries in 266 hearts of rhesus monkeys. Am. J. phys. Anthrop. *23:* 299–320 (1938).
36 CHASE, R.E. and DE GARIS, C.F.: Arteriae coronariae (cordis) in the higher primates. Am. J. phys. Anthrop. *24:* 427–448 (1939).

37 GROOVER, M.E., jr.; SELJESKOG, E.L.; HAGLIN, J.J., and HITCHCOCK, C.R.: Myocardial infarction in the Kenya baboon without demonstrable atherosclerosis. Angiology *14:* 409–416 (1963).
38 KANTARIA, P.M.: The blood supply of the heart in different monkey species. Folia primatol. *16:* 231–247 (1971).
39 GLOMSET, D.J. and BIRGE, R.F.: A morphological study of the cardiac conduction system. V. The pathogenesis of heart block and bundle branch block. Archs Path. *45:* 135–170 (1948).
40 UNGER, P.N.; GREENBLATT, M., and LEV, M.: The anatomic basis of the electrocardiographic abnormality in incomplete left bundle branch block. Am. HEART J. *76:* 486–497 (1968).
41 HILL, R.; GRESHAM, G.A., and HOWARD, A.N.: The electrocardiographic appearances of myocardial infarction in the rat. Br. J. exp. Path. *1960:* 633–637.
42 WOLF, R.H.; IEHNER, N.D.M.; MILLER, E.C., and CLARKSON, T.B.: Electrocardiogram of the squirrel monkey. J. appl. Physiol. *26:* 346–351 (1969).
43 WEISSLER, A.M.; FINGE, J., and WARREN, J.V.: The electrocardiogram of the young chimpanzee. Tech. Doc. Rept. Holloman N.M. U.S.A.F. Missile Development Centre (1961).
44 MALINOW, M.R. and DE LANNOY, C.W., jr.: The electrocardiogram of *Macaca fuscata.* Folia primatol. *7:* 284–291 (1967).
45 MALINOW, M.R.: An electrocardiographic study of *Macaca mulatta.* Folia primatol. *4:* 51–65 (1966).
46 WATT, T.B.; MURAO, S., and PRUITT, R.D.: Left axis deviation induced experimentally in a primate heart. Am. Heart J. *70:* 381–389 (1965).
47 KAMINER, B.: The electrocardiogram of the baboon *(Papio ursinus).* S. Afr. J. med. Sci. *23:* 231–240 (1958).
48 HERMANN, G.R.: The electrocardiographic patterns in 150 baboons in the domestic and African colonies at Southwest Foundation for Medical Research. Proc. 1st Int. Symp. on the Baboon and its Use as an Experimental Animal (1965).
49 CITTERS, R.L. VAN and LASRY, J.E.: Cardiovascular function in adult baboons as indicated by standard diagnostic tests. Folia primatol. *3:* 13–21 (1965).
50 CITTERS, R.L. VAN and FRANKLIN, D.L.: Cardiovascular dynamics in ambulatory baboons during spontaneous activity and exercise. Proc. 1st Int. Symp. on the Baboon and its Use as an Experimental Animal (1965).
51 STAHL, W.R. and MALINOW, M.R.: A survey of physiological measurements in *Macaca mulatta.* Folia primatol. *7:* 12–23 (1967).
52 GARDNER, M.V.: The blood picture of normal laboratory animals. A review of the literature (1936–1946). J. Franklin Inst. *244:* 151–161 (1954).
53 MELVILLE, G.S., jr.; WHITCOMB, W.H., and MARTINEZ, R.S.: Haematology of the *Macaca mulatta* monkey. Lab. Anim. Care *17:* 189–198 (1967).
54 ALLEN, J.R. and CARSTENS, L.A.: Haematologic alterations observed in newly acquired monkeys during the period of their isolation. Lab. Anim. Care *15:* 103–110 (1965).
55 FOY, H.; KONDI, A., and MBAYA, V.: Haematological and biochemical indices in the East African baboon. Blood *26:* 682–686 (1970).

56 DE LA PENA, A.; MATTHIJSSEM, C., and GOLDZIEHER, J.W.: Normal values for blood constituents of the baboon (*Papio* species). Lab. Anim. Care *20:* 251–261 (1970).
57 ANDERSON, D.R.: Normal values for clinical blood chemistry tests of the *Macaca mulatta* monkey. Am. J. vet. Res. *27:* 1484–1489 (1966).
58 ST. CLAIR, R.W.; MACNINTCH, J.E.; MIDDLETON, C.C.; CLARKSON, T.B., and LOFLAND, H.B.: Changes in serum cholesterol levels of squirrel monkeys during importation and acclimatation. Lab. Invest. *16:* 828–832 (1967).
59 ROBINSON, F.R.; GISLER, D.B., and DIXOM, D.F.: Factors influencing 'normal' S.G.O-T levels in the rhesus monkey. Lab. Anim. Care *14:* 275–282 (1964).
60 CLARKSON, T.B.; LOFLAND, H.B.; BULLOCK, B.C., and GOODMAN, H.O.: Genetic control of plasma cholesterol. Archs Path. *92:* 37–45 (1971).
61 EYESTONE, W.H.: Scientific and administrative concepts behind the establishment of primate centers. J. Am. vet. med. Ass. *147:* 1482–1487 (1965).
62 PUGATCH, E.M.J. and SAUNDERS, A.M.: A new technique for making Häutchen preparations of unfixed aortic endothelium. J. Atheroscler. Res. *8:* 735–738 (1968).
63 SHIMAMOTO, T.; YAMASHITA, Y.; NUMANO, F., and SUNAGA, T.: Scanning and transmission electron microscopic observation of endothelial cells in the normal condition and in initial stages of atherosclerosis. Acta path. jap. *21:* 93–119 (1971).
64 DAVIES, P.F.; BOWYER, D.E., and TWORT, C.H.: Scanning electron microscopy: arterial integrity after fiscation at physiological pressure. Atherosclerosis *21:* 463–469.
65 MURATA, K.: Tritiated thymidine incorporation into aortic cells *in vivo:* cell regeneration in spontaneous atherosclerosis in monkeys. Experientia *23:* 732–733 (1967).
66 Discussion Group: The proliferative nature of atherosclerosis in the artery and the process of arteriosclerosis pathogenesis; in WOLF, pp. 213–245 (Plenum Press, New York 1971).
67 CONSTANTINIDES, P. and ROBINSON, M.: Ultrastructural injury of arterial endothelium. Archs Path. *88:* 99–105 (1969).
68 CONSTANTINIDES, P. and ROBINSON, M.: Ultrastructural injury of arterial endothelium. Archs Path. *88:* 106–112 (1969).
69 CONSTANTINIDES, P. and ROBINSON, M.: Ultrastructural injury of arterial endothelium. Archs Path. *88:* 113–117 (1969).
70 CARO, C.G.; FITZ-GERALD, J.M., and SCHROTER, R.C.: Atheroma and arterial wall shear. Proc. R. Soc. Lond. B. *177:* 109–159 (1971).
71 FRANKLIN, D.E.; WATSON, N.W.; PIERSON, K.E., and CITTERS, R.L. VAN: Technique for radiotelemetry of blood-flow velocity from unrestrained animals. Am. J. med. Electr. *5:* 24–28 (1966).
72 DAVIES, D.F.: Hypothesis. An immunological view of atherogenesis. J. Atheroscler. Res. *10:* 253–259 (1969).
73 BENDITT, E.P. and BENDITT, J.M.: Evidence for a monoclonal origin of human atherosclerotic plaques. Proc. natn. Acad. Sci. USA *70:* 1753–1756 (1973).

Chapter 3

Spontaneously Occurring Atherosclerosis in Primates

The fact that most non-human primates develop atherosclerosis albeit of mild degree is regarded by many workers as a desirable feature for their use in atherosclerosis research. The erratic occurrence of lesions, however, does make the interpretation of results difficult and larger numbers of experimental animals are necessary because of this fact. For example, an extensive dissecting aneurysm of the thoracic aorta was found in an adult male free ranging howler monkey. It was one animal out of 314 studied on a field trip [1]. In this laboratory we found a similar aneurysm in an imported adult *Cebus albifrons* monkey. This unpredictable variation in the appearance of lesions has dogged atherosclerosis research from its inception. The problem has partly been resolved by the discovery of interspecies genetic differences in response to cholesterol feeding and the occurrence of lesions in such creatures as the squirrel monkey [2]. Workers in the field are now well aware of this problem. Equally important is the need to be alive to the possibility that captivity itself may alter the character and distribution of lesions independent of any experimental manipulations that might be made [3]. Unfortunately, the costs of accommodating non-human primates are nowadays so great that the number of animals available for experiment are far too small and when 4 out of 8 cynomolgus monkeys, for example, are found to have spontaneous disease affecting as much as 80% of the thoracic aortic surface it is clear that the results of any experiments done with them are likely to be meaningless.

In most non-human primates the lesions occur mainly in the aorta and to a lesser extent in the coronary arteries. These areas of atherosclerosis are usually of the fatty streak variety though some do develop gelatinous lesions and fibrous plaques. Natural as opposed to experimental occurrence of severe vascular occlusion, with or without thrombosis is a rare event [4]. Most vascular thrombi are the results of a parasitic invasion of the vessels rather than the effects of degenerative vascular disease. Natural or as some call it spontaneous atherosclerosis has been described in many primates. It is interesting that Fox [5], writing in 1933, could find only one example of arteriosclerosis amongst 204 cebus monkeys that he examined. His successor RATCLIFFE [3] found a much greater incidence of both aortic and coronary artery disease in New World monkeys. This author has recorded a steady rise in the incidence of arterial disease amongst animals in the Philadelphia Zoo and attributes this phenomenon to the effects of social pressure within the environment. RATCLIFFE emphasised the increased incidence of disease in small branches of the coronary arteries associated with perivascular fibrosis, adjacent myocardial fibrosis, and occasionally infarction. This type of disease he recorded in an 18-year-old male orang-utan and in a male chimpanzee that had been exhibited for 458 months at Philadelphia [6]. Naturally occurring arteriosclerosis has been recorded in a wide range of animals other than primates. A comparative review in 1963 emphasised that lipid was not an important component of most of these lesions [7]. The primary condition was thought to be a degeneration of elastic tissue followed by mucoid accumulation and fibrosis. Such changes occur in dog, cat, elephant and other lower vertebrates. In primates, however, lipid deposition is a usual feature and lesions are more similar to those seen in man. LINDSAY and CHAIKOFF [8] reported a study of coronary and aortic disease in 17 primate species and emphasised the similarity to human atherosclerosis. They did not see any lesions macroscopically in the coronary arteries but 33 out of 67 animals had intimal plaques in the thoracic aorta. In general, lesions were only seen in animals over 10 years old and tended to increase in frequency with age. Soft, yellow and grey fibrous plaques were observed mostly in the thoracic aorta and lesions were constantly observed in the region of the ductus scar in the thoracic aorta. Abdominal aortic involvement occurred only in the older animals and the lesions were macroscopically similar to those found in the thoracic part of the vessel.

Lesions in the coronary arteries were detected microscopically; the earliest event being fragmentation, splitting and fraying of the internal elastic lamella. This was associated with mucopolysaccharide deposition. The lesions were further elevated by duplication of elastic membranes and additional accumulation of mucosubstance in the intima. Apart from occasional fibroblasts, cells were sparse and in particular smooth muscle cells were not a feature. In even later lesions in older animals lipid appeared as well as foamy macrophages. The picture presented here by LINDSAY and CHAIKOFF is one of primary elastic damage followed by repair: collagen and reticulin also appeared in the later lesions. Lipid did not appear to play a primary role nor was the accumulation of fibrin a feature either within or on the surface of the lesions. The intramyocardial small branches of the coronary arteries were also examined and showed similar intimal thickenings to those seen in the larger coronary arteries but the occurrence of lesions in small and large vessels was not closely related. It is always difficult to know what is meant by the terms spontaneous or naturally occurring disease. It is in general true that coronary artery lesions such as LINDSAY and CHAIKOFF described are rare in monkeys and yet intimal fibroblastic proliferation with fraying of elastic lamellae can appear even in immature rhesus monkeys after as little as a month in captivity [9]. It is clearly important when considering reports of such 'spontaneous' lesions to know the conditions under which the animal is being kept.

Other workers have found different changes in the arteries of the squirrel monkey [10]. The emphasis here being more upon the appearance of intimal macrophages and smooth muscle cells with abundant cytoplasmic lipid in them. There is some debate about the origin of these cells and it may be that some of the macrophages have entered the vessel wall from the blood stream. Whatever the answer is the presence of lipid is regarded as a central early factor in atherogenesis in this species. Similar studies have been made in the baboon [11] and there are several reports confirming the same results in man [12].

The baboons that were studied by MCGILL et al. [11] were all killed and examined within a few hours of trapping and showed, as has already been described, sudanophilic lesions in their vessels. A report on two old baboons from the San Diego Zoo [13], by contrast, emphasised the paucity of lipid in the lesions and the preponderance of proliferative and elastic tissue changes. It is difficult to know what these differences mean: perhaps, as will be discussed in the next chapter, the findings illustrate the potential reversibility of the fatty lesions in the baboon and possibly in other primates including man.

In squirrel monkey, baboon and man fragmentation of elastic tissue is an early feature but this is not so in all primates. MALINOW and STORVICK [14] studied coronary artery lesions in 314 howler monkeys and found that they consisted of spindle cells, lying beneath the endothelium with only slight changes in the elastic tissue and little evidence of fibrosis. They remarked upon the presence of mucosubstances, a little stainable iron and occasional basophilia of elastic tissue which was probably due to calcification.

Intracellular and extracellular lipid was described in the aortas of 4 out of 14 chimpanzees [15]. Here again the predominant role of lipid was emphasised and the cells that were found in the thickened intima were interpreted as being of smooth muscle origin.

Morphological studies do not, at present, provide a clear view of the earliest steps in atherogenesis. Light microscopy or for that matter electron microscopy have little to offer at the moment. By multiplying the number of entities involved they have served to raise more problems than they solve. GEER et al. [16] have published elegant electron micrographs of atherosclerotic lesions produced in baboons by dietary methods over a period of 2 years. They illustrate inclusions in the endothelial cells and confirm the presence of smooth muscle cells in the lesions though these cells were not identical with smooth muscle of the adjacent media. Morphology can only tell us what is there; we urgently need to know what these cells are doing. Perhaps the solution lies elsewhere with other techniques such as tissue culture [17].

In one respect, however, a study of the nature and distribution of lesions in the vessels of different primates does help to decide mechanisms of production of lesions and the sequence of progression from the early to the later stages of atherosclerotic disease. Reference has already been made to the vexed problem of the fate of the fatty streak and baboon studies have hinted the possibility of their reversability. The other question of their subsequent development is, to some extent, answered by a study of naturally occurring, arterial disease in non-human primates. For example, recent observations of *M. irus,* the long-tailed monkey of western Malaysia, have shown a high incidence of fatty streaking in the aorta, the lesions being distributed in a way which is very like those of young human subjects [18]. The majority of the lesions of this type were found in the thoracic aorta, adjacent to intercostal artery orifices and near to the aortic valve ring. As many as 90% of these animals showed aortic fatty streaks which is a remarkably high incidence for any non-human primate; only a few showed fibrous plaques. However, the site of localisation of these plaques and the discovery of transitions between them and fatty streaks is in favour of the view that the fibrous plaque is related to the fatty streak, probably being derived from it, and is contrary to the view that the fatty streak is not an early state of an atherosclerotic lesion.

One of the main difficulties in the evaluation of so-called natural or spontaneously occurring lesions in primates is to know the antecedent history of the animals. It is well known that age, diet, stress and other factors may play important roles in determining the incidence and structure of the lesions. A study of coronary arteries of immature rhesus and vervet monkeys found no relationship between the changes in the intima and dietary or blood lipids. It was therefore postulated that stress or lack of physical activity might be important factors in the production of the lesions that were observed [9]. Intimal changes appeared in as short a time as 3–6 weeks after captivity and were even more conspicuous after 3 months. The alterations described were splitting of the internal elastic lamella and a proliferation of fibroblasts similar to those seen in young human vessels. The authors were unable to find any evidence of the musculoelastic intimal layer which develops in growing human arteries. No sex differences were observed but rhesus had more conspicuous changes than vervet monkeys. The rhesus monkeys were kept in individual cages whereas the vervets occupied a large room with the possibility of free movement; this may have been an important factor in determining the differences between the species. Several authors have been unable to relate the degree of aortic atherosclerosis to the levels of blood lipids. VAN DER WATT et al. [20] studied recently trapped baboons from two different areas in the Transvaal. The mean serum cholesterol levels for males was 104 mg% and 99 mg% for females. In this sample of animals there was no correlation between any of the lipid levels studied and the extent of aortic lesions. But the greater the weight of the animal the more extensive was the area of the aorta-bearing lesions.

One of the biggest problem in atherosclerosis research is to distinguish, if that is possible, so-called age changes from those of atherosclerosis. It is well known that the intimal thickenings or cushions that occur at divisions and branches should not be regarded as being pathological. More controversial is the relationship of diffuse intimal thickening with atherosclerosis. This well-documented condition consisting of a generalised thickening of the intima by musculoelastic and collagenous tissue is a constant feature of human arteries of older subjects [21]. Occasionally there are small droplets of lipid scattered along the elastic lamellae. Whether these changes are a stage in the atherosclerotic process or not is debatable. Suffice to say that diffuse intimal thickening is rarely mentioned in reports of non-human primate arterial disease. PRATHAP [18] described it in the abdominal aorta of 7 out of 73 macaques that he examined. VLODAVER et al. [9] state quite clearly that musculoelastic thickening of the intima as is seen in man does not develop in the young monkey. Another study of the proximal part of the left main coronary artery was made in 133 baboons ranging in age from birth to 18 years [22]. 32 of these animals were born in captivity. In 44 of the animals that had not been the subject of any experimental manipulations it was found that the arteries were normal up to the age of 7 years in males and 11 years of age in the females. That is to say that the endothelium lay directly upon the internal elastic lamella. Subsequent changes in the older animals consisted of splitting of the internal elastic lamella with lysis of some elastic fragments and the appearance of fibroblasts. Later still distinct fibrosis was observed, and the changes were similar to the early fibrotic lesions that have been described in man.

Russian workers report different results in 30 non-human primates [23]. They found age changes in the aortas of animals that were only 6 months old, the appearances being those of thin fibres splitting off the internal elastic lamella with gradual thickening of all the layers of the arterial wall. In animals of 4 or 5 years of age splitting of the elastic lamellae had proceeded further and collagen together with smooth muscle cells had appeared in the thickened intima. At the age of 7 changes were well advanced being more conspicuous in the aorta and less so in the coronary and cerebral arteries. However, it was generally true that the older the animals the more severe were the changes. This rule was not always followed for in some monkeys which were computed to be 20 years or more of age no lesions were seen at all.

The precise determination of the age of particular primates is not always easy. If the animals have been born and kept in captivity this is simple but then the problem is to eliminate the effects of captivity itself upon the occurrence of lesions that may be found. McGill et al. [11] have provided one of the largest studies on free-ranging animals thus eliminating the effects of captivity. They studied 163 baboons of all ages rising from fetuses to apparently elderly animals. They used the state of the dentition and bodily measurements in order to determine age. The youngest animal had no aortic lesions. 5 out of 40 adolescent males and 12 out of 39 adolescent females had fatty streaks in the aorta. In adults the incidence was much higher and only one adult male was an exception in that it did not have fibrous plaques in the aorta. More detailed studies on the assessment of the precise age of non-human primates is clearly needed if more exact correlations between aging and arterial disease are to be achieved. This sort of work has also been done in the rhesus and howler monkeys [24]. Malinow and Maruffo [24] assessed the factors that might be used to determine the age of 314 free-ranging howler monkeys trapped in northern Argentina. They found that body mass alone was not a reliable criterion of age particularly because of the frequently observed sexual dimorphism in non-human primates [25]. However, they were able to say that animals about 4.0 kg in weight were probably pubertal because a sudden increase of testicular weight occurred in males of that size and because pregnancy was not observed in females less than 4.0 kg weight. Other criteria for determining age was the degree of wear of the first upper molar tooth which they were able to grade and also the dry weight of the ocular lens [26]. Others have described radiological features of value in determining the age of non-human primates [27]. The results in the howler monkey show a definite increase in the degree of aortic sudanophilia with age. This relationship was more conspicuous in young male animals though sex differences in the incidence of lesions were not clearly apparent in the oldest animals examined. In many ways these findings parallel the situation in man. The results of McGill *et al*. [11] did not show any consistent association between incidence of lesions and age, sex or state of gestation, nor did Strong and Tappen [28] find any relationship to sex in *Cercopithecus* spp.

In general, it appears from all these reports that there is a definite trend of increased incidence and severity of lesions with the age of the animal but that considerable variation exists. This is one of the factors that makes experimental studies on atherogenesis in non-human primates so difficult and emphasises the important role that is played by primate centres. Here it is possible to hold large numbers of animals and also to breed strains of known age that will reduce the amount of variation that exists in animals from the wild state.

Most reports about arterial disease in non-human primates have been concerned with the state of the aorta and coronary arteries. Occasional references to cerebral vessels are made. Scherer [29] described cerebral atherosclerosis in two chimpanzees but considered it to be a rare event. In another chimpanzee Stehbens [30] found atherosclerotic lesions associated with aneurysms. Steiner *et al*. [31] found cerebral and spinal atherosclerosis in an old gorilla and Lucke [32] found cerebral atherosclerosis in 4 out of 22 monkeys that were examined. Clarkson [33] found no evidence of it in 74 squirrel monkeys nor did Strong [34] in his extensive studies on wild Kenya baboons. There is a close correlation between cerebral atherosclerosis and haemorrhage and hypertensive vascular disease in man; the fact that hypertension is a rare event in monkeys may explain the rarity of cerebral arterial lesions in these animals and incidentally also supports the important role of haemodynamic factors such as hypertension in atherogenesis [35].

One of the most extensively studied aspects of primate physiology in relation to atherosclerosis is lipid metabolism. This is largely because of the strong body of opinion that relates atherogenesis and lipid turnover in the arterial wall. The original notion was that lipids, particularly cholesterol were imbibed into the arterial wall following some injury to the endothelial lining [36]. On the whole it was a static concept that paid no attention to the fact that the artery might be capable of responding to this metabolic insult by changing its own metabolic process. Nevertheless, there is much evidence associating elevation of plasma lipids with atherosclerosis so that it is important to relate, if possible, lipid metabolism in the non-human primates with that in man if these animals are to be used for experiment.

The principal lipids in plasma are glycerides, free fatty acids, cholesterol and phospholipids. The glycerides and fatty acids are metabolic products in transit whereas cholesterol and phospholipids turn over less rapidly though cholesterol is metabolised by many tissues including the liver which degrades the sterol to bile acids. Because they are non-polar most of the lipids bind to plasma proteins. Free fatty acids bind with albumen whereas cholesterol and glycerides combine with specific apoproteins to form lipoproteins.

STEIN and STEIN [37] have provided a detailed account of the ways in which lipids are carried in plasma and of the routes of transport through the arterial wall. Cholesterol is transported via plasmalemmal vesicles into the arterial intima, the carrier molecules being low density and high density lipoproteins. They also describe the ways in which the aortic wall synthesises lipids. Using radioautography it was shown that most of this synthetic process occurs in smooth muscle cells. Another important aspect was catabolism of lipids by enzymes such as phospholipase A_2 and sphingomyelinase. It may be that such enzymes change in amount with age thus in part explaining accumulations of lipids such as sphingomyelin in the older aorta.

The main fatty acids that esterify lipids vary a good deal from one animal to another. For example linoleic acid is the principal one in man and it has been shown that it is also true for the baboon [38]. A further point of similarity is that ^{14}C-labelled linoleic acid is readily incorporated into phospholipids when slices of aortic wall taken from primates are incubated with the isotopically labelled lipid *in vitro* [39].

There are, however, certain differences between man and some other members of the primate order. Whereas the bulk of plasma cholesterol in man is carried on the β-lipoproteins or low density lipoproteins; in many other animals the α-lipoproteins or high density lipoproteins preponderate [40]. It is tempting to equate an elevated level of β-lipoproteins with a tendency to develop atherosclerosis. It has been shown that these proteins increase in baboons fed on atherogenic diets and which develop aortic atherosclerosis [41]. The situation concerning the relationship of β-lipoproteins and atherosclerosis is not, by any means, clearly defined in the primate order. For example, the marmoset is probably the least susceptible to naturally occurring atherosclerosis and yet the percentage of cholesterol carried by α- and β-lipoproteins is almost equal. The squirrel monkey has the same lipoprotein situation as the marmoset and yet is highly susceptible to the development of atherosclerosis and even though male squirrel monkeys transport a greater proportion of cholesterol on α-lipoproteins than do the females they do not seem less susceptible to atherosclerotic disease.

Studies that have been made on the extent of arterial lesions and levels of blood lipids in primates have not always confirmed the association. McGill et al. [11] examined the serum levels of cholesterol in their baboons and found levels ranging from 35 to 168 mg/100 ml with an average of 78. They did not find any tendency of serum cholesterol levels to increase with age nor did they find any relationship between the occurrence of arterial lesions and the levels of blood cholesterol. Similar studies in squirrel monkeys have shown only a small relationship between the presence of microscopically detectable aortic atherosclerosis with blood cholesterol or triglyceride levels [42]. Some workers have found that plasma lipid levels of baboons tend to decrease with age [43]. It does then appear that lesions can occur without elevation of blood lipids. This is entirely in keeping with the view that lipids aggravate but do not necessarily cause atherosclerosis to appear in the first instance. In fact, these observations strongly support the view that the basic atherosclerotic lesion is an intimal proliferation, almost a sort of response to injury and that lipids collect in them solely because of the continued bombardment of the arterial wall from high levels of plasma lipids. It may be that this is the only reason why man suffers from the occlusive complications of atherosclerosis which, in the natural state, are almost solely confined to this member of the primate order. As we shall see in the next chapter exposure of primates to high intake of dietary lipids result in the same situation that appertains in man.

Eggen et al. [44] studied cholesterol metabolism and serum lipids in young and old baboons fed upon a basal diet of chow which was low in fat content and the followed it by the same diet with the addition of saturated fats and cholesterol. They found a rise in the serum lipids in both age groups, the changes in the cholesterol fatty acids being more conspicuous in the young than in the old animals. Both groups absorbed similar amounts of cholesterol from the diet but the young animals showed greater ability to suppress their endogenous synthesis of cholesterol than did the older ones. Lofland et al. [45] have reported a similar phenomenon in the cebus monkey: older animals on high cholesterol diets maintained a higher serum cholesterol than did the younger ones but in this species they found a greater turnover rate in younger than in older animals rather than a suppression of endogenous synthesis.

There is no doubt that, in man, lipid accumulates in the arterial wall as age increases. Whether this is part of the atherosclerotic process or is an inevitable accompaniment of aging in man is difficult to say. A proper understanding of the end result namely the accumulation of lipid in the arterial wall, involves knowledge of several processes. First, the dietary absorption, transport and metabolism of these substances, second, the factors affecting levels of plasma lipids such as the interplay between various 'pools' of lipid in adipose tissue and in other organs and last, an understanding of diffusion and metabolism in the arterial wall itself. Changes in lipid content of various pools have been recorded with increasing age. It seems that the oleic acid level of adipose tissue rises with age. A similar rise in plasma cholesterol ester content is also seen in human beings with increasing age the increase being mainly in the cholesteryl linoleate fraction. This observation is, however, only applicable to developed human populations and may not be relevant to underdeveloped peoples or to other primates. Cholesteryl linoleate is mainly in the form of lipid droplets aggregated around collagen fibres in the intima; this may well be an aging change. Cholesteryl oleate, on the other hand, appears within macrophages and is probably the direct result of the atherosclerotic process [46].

The possible relationship between the degree of atherosclerosis and deposition of lipid in more accessible sites such as the eye, skin and tendons has always been of interest if only as a possible predictor of the atherosclerotic state of the individual. When serum cholesterol levels are very high as in the experimental rhesus monkey cutaneous xanthomata appear [47]. In these experiments xanthomas appeared in the skin in as short a time as 10 months after the start of cholesterol feeding. Xanthomas are also a feature of hyperlipoproteinaemia in man: it is, however, interesting to reflect that the preponderant lipid in the skin lesions is cholesteryl oleate, the amount of cholesteryl linoleate being low compared to plasma levels of that lipid so that xanthoma lipids are clearly not entirely explicable in terms of deposition from the plasma [48]. Ho and Taylor [49] studied the cholesterol content of four species of mammals after cholesterol feeding. In the rabbit and prairie dog cholesterol increased in most tissues, not so in the rat and dog where aorta and skin showed the greatest increases. If rabbits were taken off the diet it was found that most tissues cleared the excess cholesterol but that skin and aorta cleared slowly and incompletely. Here then is some evidence to relate lipid accumulation in vessels and skin [49].

BOUISSOU et al. [50] also correlated the degree of aortic atherosclerosis at post-mortem examination with the total extractable sterol level in pieces of skin taken from the body. Bearing in mind the high cost of non-human primates for research such methods of assessing atherosclerosis during the course of experiments may prove useful.

Another correlate of a tissue lipid level with that in arteries is provided by studies on human arteries, tendon and fascia [51]. These workers found lipid deposits in tendons as early as the age of 15 years. These lipids were cholesteryl esters predominately. It was interesting that the lipid level in the Achilles tendon did not correlate well with deposition in the aorta but did relate to that in the coronary arteries.

In this context also it is often said that arcus senilis in man is related to hypercholesterolaemia. I have only seen this condition in the rabbit fed cholesterol and that on a few occasions only. It is not recorded as occurring in non-human primates under conditions of cholesterol feeding, or under natural conditions.

It is remarkable how rarely are reported examples of myocardial infarction and vascular thrombosis in association with atherosclerosis of non-human primates. Very few examples of recent or old myocardial necrosis have been recorded. MANNING [52] described myocardial infarction in an 8-year-old female chimpanzee; RATCLIFFE et al. [53] recorded coronary occlusion with adjacent fibrosis in an 18-year-old male orang-utan and TAYLOR et al. [54] produced myocardial infarction in M.mulatta by diet induced hypercholesterolaemia. Rarely myocardial infarction has been recorded in a primate where no demonstrable atherosclerosis was present. GROOVER et al. [55] reported an example in a Kenya baboon: the possibility of small platelet emboli as a factor cannot be excluded.

It is also equally remarkable that there are few records of thrombi in atherosclerotic vessels of non-human primates. This may be partly because it is easy to detach and overlook small thrombi if arteries are opened before they are examined histologically. However, a histological examination of the previously opened arteries of captured wild M.mulatta showed small thrombi in the aorta, pulmonary artery and carotids in 5 out of 100 animals [56]. RATCLIFFE [57] found a thrombosed atherosclerotic abdominal aorta in an old chimpanzee out of 93 non-human primates examined at the Philadelphia Zoo and VASTESAEGER et al. [58] described coronary thrombosis in one chimpanzee. One case of occlusive thrombosis of the main pulmonary artery in a 4-year-old M.mulatta has been described in 1968 [59].

The problem of thrombosis is a paramount one in the elucidation of the causes of human coronary occlusion and it is disappointing that so little has been reported or has been done to study the matter in non-human primates. Some work has already been done to study the basic parameters such as platelets, coagulation factors and fibrinolysis in non-human primates as compared to man. This is shown by an increased tendency to develop contact activation to factors XI and XII. In addition there is also greater activity of factors II and VII. It seems also that platelet counts in the blood of non-human primates are higher than those of man. Platelets in these animals are being studied for there are several aspects of platelet physiology such as adhesion, aggregation and dissolution that play a part in thrombus formation. MILLS [62] has shown a biphasic response of baboon platelets to the addition of ADP in the aggregometer. This type of aggregation is potentiated by epinephrine and because this phenomenon is a feature of primates it may well be an important factor to consider in the relationship between stress and atherosclerotic disease.

Studies on non-human primates that we have discussed so far would suggest that far from being less liable to produce thrombi non-human primates seem to be more prone than man. However, there is another side of the problem of thrombosis to consider, namely fibrin formation and fibrin dissolution. Non-human primates have lower fibrinogen and higher plasminogen levels than man: perhaps these results explain the different thrombogenic tendencies in the primate order.

Several other diseases of the cardiovascular system have been described in non-human primates. In many the cause is obvious and the lesions should not be confused with atherosclerosis or ischaemic myocardial disease. 20 recently imported rhesus monkeys from India were examined and 18 showed either focal or confluent myocarditis [63]. No cause for this could be found nor were the SGOT or SGPT levels elevated in the blood. Toxoplasmosis is not infrequently found particularly in New World monkeys and this may produce myocardial damage the cause of which may not always be obvious because the protozoa are not seen in the lesions in all cases [64].

A good deal of confusion has also been created in the literature of times past by the view that primary medial arterial changes are part of the atherosclerotic process. It is important that conditions such as mediocalcinosis (fig. 5) which can occur in many animals, rabbits, cattle, rats, non-human primates, etc., should be distinguished from atherosclerosis. MARUFFO and MALINOW [65] make this point quite clear in a report of mediocalcinosis in the howler monkey. They describe the changes in 1 animal out of 314 examin-

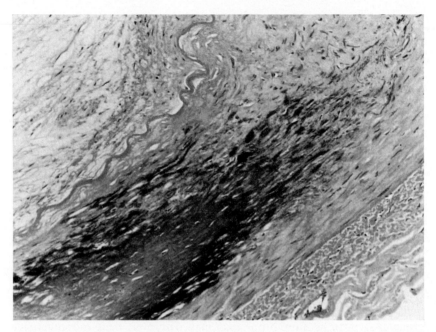

Fig. 5. Calcification of the arterial media. This mediocalcinosis can readily be produced by a dietary excess of vitamin D_3. He. × 260.

ed on a field trip in northern Argentina. The entire aorta was involved by medial calcification adjacent to which was abundant neutral polysaccharide material. They thought that the process was not caused by hypervitaminosis D_3 because of the regular distribution of calcium in the aortic wall. No intimal sudanophilia accompanied these lesions. On the other hand, the spontaneous coronary lesions that MALINOW and STORVICK [14] described in howler monkeys could well be regarded as early atherosclerosis. These lesions were segmental consisting of intimal cell proliferation, accumulation of mucopolysaccharide with calcification of the adjacent internal elastic lamella. In human terms they could be regarded as a mature variant of the gelatinous plaque [66].

Finally, it is important not to err by regarding any differences that may be observed in the structure of arteries of non-human primates from those of man as being pathological. For example, the small pulmonary arteries of *Cercopithecus* spp. were found to be unusually muscular being some 5–10% thicker than vessels of comparable size in the human being. This was regarded as abnormal. It seems likely that it is a normal variation, that cannot be correlated with any other abnormality [67]. Similar thick walled pulmonary vessels occur in pig and rabbit.

It is, I suppose, reasonable to expect, at the end of a chapter on spontaneous atherosclerosis, to find some comment about any indications that might point to one species being more suitable than another for use in experimental atherosclerosis research. Many of those who work in this field have debated this matter for years and in general despite much discussion each worker tends to carry on with the same sort of monkey that he selected at the start of his work. Any answer to this problem is so weighted with imponderables as to be of little value. If availability, cost of purchase, cost of housing and maintenance can be excluded from the discussion and if breeding facilities are freely available then the tendency would be, I suspect, to select the animal that freely develops spontaneous atherosclerosis. The prevalence and severity of atherosclerotic disease amongst the primates is indicated in a paper by STRONG and TAPPEN [28]. Baboon, chimpanzee [68] and spider monkey head the list. At the bottom is the marmoset that resembles the rhesus, cebus and red-tail monkey in having few atherosclerotic lesions under natural circumstances. Not included in that list is the long-tailed Malaysian monkey *(Macaca irus)* [18]: at the present time it fulfills all the criteria of desirability as an experimental model.

References

1 MARUFFO, C.A. and MALINOW, M.R.: Dissecting aneurysm of the aorta in a howler monkey *(Alouatta carya)*. J. Path. Bact. *92:* 567–570 (1966).
2 CLARKSON, T.B.; LOFLAND, H.B.; BULLOCK, B.C., and GOODMAN, H.O.: Genetic control of plasma cholesterol. Archs Path. *92:* 37–45 (1971).
3 RATCLIFFE, H.L.; YERASIMIDES, T.G., and ELLIOT, G.A.: Changes in the character and location of arterial lesions in mammals and birds in Philadelphia zoological garden. Circulation *21:* 730–738 (1960).
4 CHAKRAVARTY, R.N. and CHAWLA, K.K.: Spontaneously occurring mural thrombi in arteries of *Macaca mulatta*. J. Atheroscler. Res. *6:* 455–462 (1966).
5 FOX, H.: Arteriosclerosis in lower mammals and birds; in COWDRY Arteriosclerosis: a survey of the problem, p. 153 (Macmillan, New York 1933).
6 RATCLIFFE, H.L. and SYNDER, R.L.: Atherosclerosis in mammals and birds at the Philadelphia Zoo; in ROBERTS and STRAUS Comparative atherosclerosis, pp. 127–128 (Harper & Row, New York 1965).
7 LINDSAY, S. and CHAIKOFF, I.L.: Naturally occurring arteriosclerosis in animals. A comparison with experimentally induced lesions; in SANDLER and BOURNE Atherosclerosis and its origin, pp. 349–437 (Academic Press, New York 1963).

8   LINDSAY, S. and CHAIKOFF, I.L.: Naturally occurring arteriosclerosis in non-human primates. J. Atheroscler. Res. 6: 36–61 (1966).
9   VLODAVER, Z.; MEDALIE, J., and NEUFELD, H.N.: Coronary arteries in immature monkeys. J. Atheroscler. Res. 8: 923–933 (1968).
10 MIDDLETON, C.C.; CLARKSON, T.B.; LOFLAND, H.B., and PRICHARD, R.W.: Atherosclerosis in the squirrel monkey. Archs Path. 78: 16–23 (1964).
11 MCGILL, H.C., jr.; STRONG, J.P.; HOLMAN, R.L., and WERTHESSEN, N.T.: Arterial lesions in the Kenya baboon. Circulation Res. 8: 670–679 (1960).
12 HOLMAN, R.L.; MCGILL, H.C., jr.; STRONG, J.P., and GEER, J.C.: Natural history of atherosclerosis. Early aortic lesions as seen in New Orleans in the middle of the 20th century. Am. J. Path. 34: 209–235 (1958).
13 LINDSAY, S. and CHAIKOFF, I.L.: Arteriosclerosis in the baboon. Archs Path. 63: 460–471 (1957).
14 MALINOW, M.R. and STORVICK, C.A.: Spontaneous coronary lesions in howler monkeys (Alouatta carya). J. Atheroscler. Res. 8: 421–431 (1968).
15 STRONG, J.P.; KRITCHEVSKY, D., and MARTINEZ, R.D.: Sudanophilic aortic lesions in chimpanzees. Archs Path. 81: 544–547 (1966).
16 GEER, J.C.; CATSULIS, C.; MCGILL, H.C., jr., and STRONG, J.P.: Fine structure of the baboon aortic fatty streak. Am. J. Path. 52: 265–286 (1968).
17 POLLAK, O.J.: Monographs on atherosclerosis, vol. 1 (Karger, Basel 1969).
18 PRATHAP, K.: Spontaneous aortic lesions in wild adult long-tailed monkeys (Macaca irus). J. Path. Bact. 110: 135–143.
19 MITCHELL, J.R.A. and SCHWARTZ, C.J.: in Arterial disease, p. 45 (University Press Oxford 1965).
20 WATT, J.J. VAN DER; KOTZE, J.P.; KEMPFF, P.G.; DU PLESSIS, J.P., and LAUBSCHER, N.F.: Aortic intimal lesions and serum lipids in wild baboons. J. med. Primatol. 2: 25–38 (1973).
21 MOVAT, H.Z.; MORE, R.H., and HAUST, M.D.: The diffuse intimal thickening of the human aorta with ageing. Am. J. Path. 34: 1023–1031 (1958).
22 GILBERT, C. and GILLMAN, J.: Structural modification in the coronary artery of the baboon (Papio ursinus) with special reference to age and endocrine status. S. Afr. J. med. Sci. 25: 59–70 (1960).
23 LAPIN, B.A. and YAKOVLEVA: in WINDLE Comparative pathology in monkeys, p. 132 (Thomas, Springfield 1936).
24 MALINOW, M.R. and MARUFFO, C.A.: Naturally occurring atherosclerosis in howler monkeys (Alouatta carya). J. Atheroscler. Res. 6: 368–380 (1966).
25 WAGENEM, G. VAN and CATCHPOLE, H.R.: Cervical growth of the rhesus monkey (Macaca mulatta). Am. J. phys. Anthrop. 14: 245–273 (1956).
26 LORD, R.D., jr.: The lens as an indicator of age in cotton tail rabbits. J. Wild. Mgmt. 23: 358–365 (1959).
27 HAIGH, M.V. and SCOTT, A.: Some radiological and other factors for assessing age in the rhesus monkey using animals of known age. Lab. Anim. Care 15: 57–73 (1965).
28 STRONG, J.P. and TAPPEN, N.C.: Naturally occurring arterial lesions in African monkeys. Archs Path. 79: 199–205 (1965).
29 SCHERER, H.J.: Vergleichende Pathologie des Nervensystems der Säugetiere unter besonderer Berücksichtigung der Primaten (Thieme, Leipzig 1944).

30 STEHBENS, W.E.: Cerebral aneurysm in animals other than man. J. Path. Bact. *86:* 160–168 (1963).
31 STEINER, P.E.; RASMUSSEN, T.B., and FISCHER, L.E.: Neuropathy, cardiopathy, haemosiderosis and testicular atrophy in *Gorilla gorilla*. Archs Path. *59:* 5–25 (1955).
32 LUCKE, R.: Spontaneous cerebral lesions in monkeys. Archs Neurol., Chicago *10:* 212–225 (1923).
33 CLARKSON, T.B.: Spontaneous atherosclerosis in subhuman primates; in ROBERTS and STRAUS Comparative atherosclerosis, pp. 211–214 (Hoeber, New York 1965).
34 STRONG, J.P.: Arterial lesions in primates; in ROBERTS and STRAUS Comparative atherosclerosis, pp. 244–252 (Hoeber, New York 1965).
35 LUGINBÜHL, H.: Vascular disease in animals. Comparative aspects of cerebrovascular anatomy and pathology in different species; in Cerebral vascular diseases, pp. 3–27 (Grune & Stratton, New York 1966).
36 MARCHAND, F.: Über Arteriosklerose. Verh. Kongr. inn. Med. *21:* 23–59 (1904).
37 STEIN, Y. and STEIN, O.: Lipid synthesis and degradation and lipoprotein transport in mammalian aorta; in Atherogenesis: initiating factors, pp. 165–183 (Elsevier/North-Holland, Amsterdam 1973).
38 KRITCHEVSKY, D.; SHAPIRO, I.L., and WERTHESSEN, N.T.: Biosynthesis of cholesterol in the baboon. Biochim. biophys. Acta *65:* 556–557 (1962).
39 STEIN, Y. and STEIN, O.: Incorporation of fatty acids into lipids of aortic slices of rabbits, dogs, rats and baboons. J. Atheroscler. Res. *2:* 400–412 (1962).
40 VASTESAEGER, M.M. and DELCOURT, R.: The natural history of atherosclerosis. Circulation *26:* 841–855 (1962).
41 GRESHAM, G.A. and HOWARD, A.N.: Vascular lesions in primates. Ann. N.Y. Acad. Sci. *127:* 694–701 (1965).
42 LOFLAND, H.B.; ST. CLAIR, R.W.; MACNINTCH, J.E., and PRICHARD, R.W.: Atherosclerosis in New World primates. Archs Path. *83:* 211–214 (1967).
43 ZYL, A. VAN and KERRICK, J.E.: The serum lipids and age in the baboon *(Papio ursinus)*. S. Afr. J. med. Sci. *20:* 97–117 (1955).
44 EGGEN, D.A.; NEWMAN, W.P., and STRONG, J.P.: Absorbtion and turnover of cholesterol in the baboon; in VATBORG The baboon in medical research; 2nd ed., pp. 559–569 (University of Texas Press, Austin 1965).
45 LOFLAND, H.B.; CLARKSON, T.B.; ST. CLAIR, R.W.; LEHNER, N.D.M., and BULLOCK, B.C.: Atherosclerosis in *Cebus albifrons* monkeys. 1. Sterol metabolism. Expl. molec. Path. *8:* 302–313 (1968).
46 KRITCHEVSKY, D.: Cholesterol metabolism in aorta and tissue culture. Lipids *7:* 305–309 (1972).
47 ARMSTRONG, M.L.; CONNER, W.E., and WARNER, E.D.: Xanthomatosis in rhesus monkeys fed a hypercholesterolemic diet. Archs Path. *84:* 227–237 (1967).
48 PARKER, F. and SHORT, J.M.: Xanthomatosis associated with hyperlipoproteinemia. J. invest. Derm. *55:* 71–88 (1970).
49 HO, K.J. and TAYLOR, C.B.: Comparative studies on tissue cholesterol. Archs Path. *86:* 585–596 (1968).
50 BOUISSOU, H.; PIERAGGI, M.T.; JULIAN, M.; CUSACIL, I.; DOUSTE-BLAZY, L.; LATORRE, E., and CHARLET, J.P.: Identifying arteriosclerosis and aortic atheromatosis by skin biopsy. Atherosclerosis *19:* 449–458 (1974).

51 ADAMS, C.W.M.; BAYLISS, O.B.; BAKER, R.W.R.; ABDULLA, Y.H., and HUNTER-CRAIG, C.J.: Lipid deposits in aging human arteries, tendons and fascia. Atherosclerosis *19:* 429–440 (1974).
52 MANNING, G.W.: Coronary disease in the ape. Am. Heart J. *23:* 719–724 (1942).
53 RATCLIFFE, H.L.; YERASIMIDES, T.G., and ELLIOT, G.A.: Changes in the character location of arterial lesions in mammals and birds in the Philadelphia zoological garden. Circulation *21:* 730–738 (1960).
54 TAYLOR, C.B.; MALNALO-ESTRELLA, P. and COX, G.E.: Atherosclerosis in rhesus monkeys. Archs Path. *76:* 239–249 (1963).
55 GROOVER, M.E., jr.; SELJESKOG, E.L.; HAGLIN, J.J., and HITCHCOCK, C.R.: Myocardial infarction in the Kenya baboon without demonstrable atherosclerosis. Angiology *14:* 409–416 (1963).
56 CHAKRAVARTY, R.M. and CHAWLA, A.K.: Spontaneously occurring mural thrombi in arteries of *Macaca mulatta*. J. Atheroscler. Res. *6:* 455–462 (1966).
57 RATCLIFFE, H.L.: Age and environment as factors in the nature and frequency of cardiovascular lesions in mammals and birds in the Philadelphia zoological garden. Ann. N.Y. Acad. Sci. *127:* 715–735 (1965).
58 VASTESAEGER, M.M.; GILLOT, P. et PARMENTIER, R.: L'athérosclérose coronarienne chez les vertébrés supérieurs vivant en jardin zoologique. Acta cardiol. *15:* 12–30 (1960).
59 ULLAND, B.M.: Chronic occlusive thrombosis of the pulmonary trunk and main right pulmonary artery in a four year old *Macaca mulatta*. Br. vet. J. *124:* 245–247 (1968).
60 SEAMAN, A.J. and MALINOW, M.R.: Blood clotting in non-human primates. Lab. Anim. Care *18:* 80–84 (1968).
61 MACFARLANE, R.G.: in The homeostatic mechanism in man and other animals (Academic Press, New York 1970).
62 MILLS, D.C.B.: in The homeostatic mechanism in man and other animals (Academic Press, New York 1970).
63 SOTO, P.J.; BEALL, F.A.; NAKAMURA, R.M., and KUPFERBERG, L.L.: Myocarditis in rhesus monkeys. Archs Path. *78:* 681–690 (1964).
64 MCKISSICK, G.E.; RATCLIFFE, H.L., and KOESTNER, A.: Enzootic toxoplasmosis in caged squirrel monkeys. Path. Vet. *5:* 538–560 (1968).
65 MARUFFO, C.A. and MALINOW, M.R.: Aortic mediocalcinosis in a howler monkey. J. Path. Bact. *92:* 236–238 (1966).
66 GEER, J.C. and HAUST, D.M.: Smooth muscle cells in atherosclerosis. Monogr. Atheroscler., vol. 2, pp. 61–62 (Karger, Basel 1972).
67 JONES, E.L.: Pulmonary arteries in vervet monkeys. J. Path. Bact. *99:* 181–191 (1969).
68 BOURNE, G.H. and SANDLER, M.: Atherosclerosis in chimpanzees; in The chimpanzee, vol. 6, pp. 248–264 (Karger, Basel/University Park Press, Baltimore 1972).

Chapter 4

Experimental Atherosclerosis

Introduction

Experimental animals have been used in a wide range of research in the biomedical field. For most kinds of work they are accepted as being suitable models for the potential solution of problems but their use in atherosclerosis research has always been hotly debated. This is largely because the lesions produced rarely approximate in appearances to the late stage of atherosclerosis as observed in man. It is well recognised that little knowledge can be gained about the pathogenesis of glomerulonephritis from a study of severely scarred, shrunken kidneys but this obvious conclusion has not, in times past, been applied to a study of experimental atherosclerosis. Gradually, however, there has been acceptance of the need for an attack upon the problem of early lesions and some have even proposed that study should not be of early morphological changes but of those biochemical disorders that precede them. Fatty streaks in non-human primates are now acceptable models of human atherosclerosis and it is the primary purpose of this chapter to discuss the various ways of eliciting their formation and also to consider the significance of the results that have been achieved so far, in relation to an understanding of human atherogenesis.

Most workers adhere to the use of a particular species of non-human primate largely because of familiarity with its behaviour, habits and variability of the particular animals that they use. It matters little which animal is employed since they all have something to teach us because experiments with them often raise more unsolved problems than they do provide solutions. Most workers in this field would agree that the object is to produce and to study changes in the intima of arteries and preferably in the intima of coronary arteries. Broadly speaking three sorts of lesions tend to appear in various experiments. There are, first the lipid-rich lesions analogous to those produced by cholesterol feeding in the rabbit; these have features in common with the human fatty streak. Second there are proliferative, intimal lesions where smooth muscle cells, connective tissue fibres and mucopolysaccharide may be found. This type of lesion is similar to any reparative process to be seen following tissue injury in many organs of the body. Third there is subendothelial oedema which is an early change probably representing a disturbance of fluid transport in the intima. This may be the first visible step in the atherosclerotic process though it is important to remember and to investigate metabolic events that may precede even this visible sign of intimal injury.

Earlier experiments with non-human primates were designed to investigate those risk factors that were known to be associated with atherosclerosis in man. The notion that diet and in particular fatty diets are important in atherogenesis has provided the basis for the design of many experiments. Cholesterol feeding has been a central factor in most of these. Other factors such as hypertension, stress, obesity, the effect of lack of exercise, smoking and associated hypoxia, sexual differences, the effects of hormones and the association with diabetes mellitus have only received passing mention from time to time. Experiments concerned with thrombogenesis and the subsequent fate of thrombi are few in the non-human primate.

The possibility that lesions, once produced, might be reversible has only been considered recently. Before 1960 there were few experiments to do with this matter. Clearly this is an important topic if we are to do anything about established atherosclerotic lesions in man. As we shall see later, lesions are reversible but only in that the lipid component can be reduced by suitable manipulation of the diet: the connective tissue component of the lesion remains. The results tend to confirm the view that the atherosclerotic lesion is initially a response to injury and that other components of it such as lipids, thrombi and so on are secondary events that are superimposed on the primary 'inflammatory' event. With few exceptions this view of a response to injury dominates the experimental approaches to a study of atherosclerosis in non-human primates. For example the thrombogenic hypothesis is considered unlikely to be tenable as originally proposed though the possibility that platelets might provide a means of endothelial injury and subsequent development of atherosclerosis cannot be readily discarded. I have held this view for some years and it is becoming increasingly supported by recent work on the implication of the platelet in the inflammatory response much of which is being done by GORDON and his associates here in Cambridge.

This chapter will deal with the various ways in which atherosclerosis has been provoked in the non-human primate and with the morphological and biochemical aspects of these experimental animals. Of all creatures the non-human primate is most sensitive to situations of housing, size of animal groups, methods of handling and so on. The effect of these, often ill defined, variables must always be considered when trying to interpret the results of the experiments that follow.

Dietary Experiments

Surprisingly little work was done with non-human primates before the 1950s. GOLDBLATT's [1] work with macaques when he succeeded in producing hypertension by constricting renal arteries was one of the few recorded successful results. Others attempted to produce atherosclerosis by feeding cholesterol and failed [2]. In these experiments rhesus monkeys were used and were considered to be resistant to the production of cholesterol-induced atherosclerosis: it was not in fact until 20 years later that TAYLOR *et al.* [3] succeeded in doing so. Before this, SPERRY *et al.* [4] tried to augment the effects of cholesterol feeding by thyroidectomy but had not managed to do more than produce a slight elevation in the serum cholesterol. HUEPER in 1946 fed 1% cholesterol to rhesus monkeys for 18 months with little success. It is not surprising that non-human primates, which have many attractions to the experimenter because of morphologic and metabolic similarities to man, were little used until the early 1950s when the classic experiments of TAYLOR *et al.* [3] set the pace for future work. Since that time, however, the going has not been easy and results have often been confusing. An important factor which was not appreciated early on was the variability of the interspecies response of non-human primates to dietary manipulation. This was demonstrated by LOFLAND *et al.* [5] who coined the terms hyporesponder and hyperresponder to describe the phenomenon in squirrel monkeys fed on cholesterol (0.5 mg/kg). STRONG and McGILL [6] likewise observed a marked variation in response of baboons fed on identical diets.

In 1952, TAYLOR and his colleagues began an evaluation of the rhesus monkey as a model for atherosclerosis research. It was shown that when these animals were fed diets containing concentrations of fat and cholesterol comparable to those generally ingested by people in the United States that the animals developed a moderate hypercholesterolaemia. After a few months on this sort of regime the animals were also found to have atherosclerosis [7]. In brief, these workers fed five different isocaloric diets which contained 30% of the calories as lipids together with 3 g of cholesterol. The blood cholesterol levels rose and after 4 years of feeding cutaneous xanthomas were produced and one animal developed myocardial infarction. In the latter, a female monkey, there was severe coronary atherosclerosis and superimposed thrombosis [8].

In a subsequent paper they described the evolution and progression of the lesions [9] which were similar to those of man. The earliest lesions appeared in the aorta, iliac arteries and carotid sinus. Almost as rapidly, lesions appeared in the proximal parts of the coronary arteries. If the period of hypercholesterolaemia was prolonged lesions developed in more peripheral vessels such as small branches of the coronary arteries, mesenteric, splenic and femoral arteries and in the arteries of the circle of Willis.

The early lesions, like those of man, were primarily intimal consisting of diffuse fine deposits of interstitial lipid. After almost 9 months, lipophages had collected and local mesenchymal cells had been stimulated to produce an intimal scar. It needed 36–65 months of feeding to produce fragmentation of the internal elastic lamina with deposits of lipid in the adjacent media. They did not find any significant difference in response between female and male animals in this study.

A similar experiment on a single adult rhesus monkey given a diet enriched with cholesterol and fat showed essentially similar results. The animal developed arterial lesions and xanthomatosis. In addition, the presence of a preponderant β-lipoproteinaemia was reported [10], and this is another aspect in which the rhesus model approximates to the human atherosclerotic subject. As many as 86% of rhesus monkeys observed by ARMSTRONG et al. [11] developed xanthomatosis on a raised cholesterol intake. The lesions tended to be more widely distributed than those of man and it is interesting that the animals did not develop xanthelasma or corneal arcus. In this experiment also there was an elevation of low density lipoproteins in the S_f 0–20 range. They did not show any hypertriglyceridaemia a situation again similar to that in man. In this particular primate study xanthomatosis was the external manifestation of hypercholesterolaemia. The lesions were distributed in a way similar to those found in man. It was, however, interesting that they had not produced xanthelasma nor corneal arcus both of which are exclusively human manifestations. This particular experimental model was an example of diet-induced β-hyperlipoproteinaemia without concomitant hypertriglyceridaemia and as such is of value for the study of one single lipid factor in atherogenesis.

A detailed study by ARMSTRONG and WARNER showed a different distribution of lesions in the hypercholesterolaemic rhesus monkey from those found by TAYLOR et al. [9]. TAYLOR emphasised the resemblance of the rhesus lesions to those found in man. ARMSTRONG, however, thought that the lesions in their animals were more like those of hyperlipidaemic human subjects. In his series the vessels involved were principally the proximal cervical vessels and the distal vessels supplying the limbs. The aorta was conspicuously not involved. The explanation for those differences may have been that ARMSTRONG's animals were older than those of TAYLOR at the start of the experiment. The lesions varied from minor degrees of lipid infiltration to massive intimal thickening coupled with medial lipogranulomatous arteritis with occasional giant cells. A conspicuous cellular response consisting mainly of lymphocytes was a feature of these lesions and is often seen in other non-human primates where atherosclerotic lesions have been induced by hypercholesterolaemia. It may be that this cellular picture represents an auto-allergic mechanism in the vessel wall in response to damage of medial elements such as smooth muscle cells [13].

MANNING and CLARKSON [14] obtained results that were similar to those described by TAYLOR. Again they used young male rhesus monkeys the average weight being about 3.8 kg. Fatty streaks appeared in the extramural coronary arteries and in the thoracic aorta in as short a time as 4–6 months. However, it required 18 months feeding of 1 mg cholesterol/kcal before raised lesions appear in the abdominal aorta. In this animal, as with *M. irus,* lesions appeared in the coronary arteries before they could be detected in the aorta making these non-human primates valuable models for the study of early induced coronary artery disease. Histologically, the lesions were mainly composed of lipid-laden cells and only occasionally were the more proliferative types of lesion seen similar to those described by SCOTT *et al.* [15]. The latter were composed mainly of spindle-shaped cells many of smooth muscle type. The explanation for this difference may have been in the use, by SCOTT *et al.* of peanut oil to induce lesions: this has been shown to be a powerful inducer of an intimal proliferative response by GRESHAM and HOWARD [16].

More recently, VESSELINOVITCH *et al.* [17] have reiterated the atherogenic potency of peanut oil. They fed three groups of rhesus monkeys on diets containing 2% cholesterol and 25% lipid. The lipid was corn oil, butter fat or peanut oil in the different groups. Butter fat produced the highest serum cholesterol levels and severe aortic lesions composed largely of lipid with little cellular or collagenous proliferation. Peanut oil, on the other hand, produced severe proliferative atherosclerosis of the aorta and coronary arteries with intimal cellular proliferation and fibrosis. In this group also severe coronary artery narrowing occurred. The corn oil diet induced scanty lesions compared with the other two regimes. It is clear from these studies that the type, severity and distribution of lesions will vary considerably with the type of fat that is fed in experimental diets.

Scott et al. [15] emphasise that all gradations can be found from a primitive precursor through to the mature smooth muscle cell in the intimal lesions, many of the cells being similar to those described in early human lesions [18, 19]. Another interesting feature of the lesions that they observed was necrosis of the myointimal cells and they wondered why this had occurred. No doubt it was partly due to the atherogenic stimulus that they had used, namely peanut oil. The other point that they raised was the possibility that continued necrosis of cells might provide a stimulus for sustained proliferation of smooth muscle cells that made up the bulk of the lesion. It is likely that a similar state of affairs exists in man where the progressive insudation of β-lipoprotein into the inner arterial wall leads to gradual death of smooth muscle cells and consequent proliferative replacement. In this respect the peanut oil model is an accurate reflection of the state of affairs in the progressively atherosclerotic human vessel.

Other points of importance arise from the work of Manning and Clarkson [14]. We have already referred to the difference in time of onset of thoracic and abdominal aortic lesions. The fact that fatty streaks were less frequent in the thoracic aorta of animals fed for 18 months as compared with those fed for only 4 months suggests that fatty streaking may regress. This is an important point that we proposed in a previous chapter and has also been suggested by McGill [20]. They also showed that the threshold of free cholesterol concentration required before lesions were visible was of the order of 7–8 mg/g of tissue. Normal aortas from control animals have a concentration of about 2 mg/g of wet tissue. This observation illustrates that the aorta has considerable ability to accumulate cholesterol before visible lesions occur and emphasises the importance of studies of the early biochemical events in atherogenesis.

Perhaps the most significant point they make is that coronary lesions occur before aortic lesions as in *M. irus* and that the latter animal is much less liable than the rhesus monkey to develop aortic disease. This illustrates an important feature that occasionally appears in human disease namely the presence of severe, patchy occlusive coronary disease in young men with little aortic involvement. It is likely in these cases that a genetic mechanism is responsible. *M. irus* responds well to a diet containing 1.4% cholesterol by weight [20]; all the experimental animals developed serum cholesterol greater than 350 mg/100 ml and also had severe atherosclerosis of the proximal coronary arteries which at some points reduced the size of the lumen by 50%. The small intramural branches were not involved the appearances being similar to those in man where small vessel disease is not present unless there is concomitant hypertension [22]. Aortic lesions in *M.irus* developed only after a relatively long period of dietary treatment. Of all the arteries in this animal, as in many other non-human primates, the cerebral vessels seem to be immune to the atherosclerotic process. Histologically the coronary lesions were either of proliferative type with mucopolysaccharide and little lipid or were lipid rich. A feature of all the lesions was destruction of the elastic membranes. It was important to note that half of the control animals had microscopically detectable lesions in the coronary arteries consisting of elastic tissue fragmentation and mucopolysaccharide accumulations. Lipid-rich lesions, however, were seen only in the experimental animals.

These small intimal non-lipid lesions are similar to those that have been variously described in human coronary arteries in subjects as young as 5 years of age [23]. We have already referred to the notion that these small lesions are the substrate for the subsequent development of lipid-containing atherosclerotic lesions. It is probable that diet may be a factor in man that causes progression of these small lesions to larger, lipid-rich occluding masses. In favour of this view is the frequent occurrence of small coronary lesions in the infants of Yemenite Jews [24] in whom more advanced atherosclerosis rarely develops later in life unless the diet is changed to one with a higher cholesterol content [25]. It is clear that *M.irus* is unusually susceptible to high cholesterol in the diet and this is supported by the view that three quarters of the serum cholesterol was carried by β-lipoprotein in these experimental animals. LEMAIRE and COTTET [26] have proposed the view that the susceptibility of various species to atherosclerosis is related to the ratio of α- and β-lipoproteins in the serum. In species susceptible to atherosclerosis, including man, the ratio of α- to β-lipoprotein cholesterol is usually much lower than in species resistant to the disease.

It does seem then that *M.irus* is probably the most useful model that we have amongst the non-human primates for the study of coronary artery atherogenesis and that there is no need for heroic measures such as feeding 10% cholesterol together with ^{131}I as has been done with rhesus monkeys in order to secure an adequate model [27].

Now to turn to another animal, the baboon, that has found considerable popularity in this field of research. These primates are more amenable and less liable to infections that are a danger to man than is the rhesus monkey.

One of the problems of research using this animal is the difficulty of precise species identification which is not as easy as many would have us believe. In addition, even within a species there are differences in the response of one animal and another in regard to the hypercholesterolaemic response to dietary stimulation. Nevertheless, the baboon has many features to recommend it for research into the problems of human atherogenesis. It has many features, both biochemical and structural, that resemble those of man [28], such as cholesterol and other lipid metabolism [29] blood groups and anatomy [27]. Like man baboons develop atherosclerosis under 'natural' conditions and the pattern of distribution of lesions, affecting principally the abdominal aorta is most like that seen in man.

Experiments with baboons have been made comparatively recently but already the literature is immense. As with so many non-human primates, one of the difficulties is to assess the effect of captivity upon the experimental animals. For example, wild baboons have a lower plasma cholesterol than those that have been kept for some time in captivity [30]. A similar phenomenon has, already, been described in the squirrel monkey (q.v.). We have already discussed the potential effects of methods of housing on the development of arterial diseases in captive animals. Other variables change when baboons are brought in from the wild state. For example the haemoglobin level of the blood rises [31], probably due to the treatment for parasitic infestation. But as we shall see later hypoxia may well be a factor in atherogenesis which means that we cannot safely use recently captured wild baboons as controls for experimental animals because the arterial intima of such animals is already exposed to the chance of hypoxic damage from anaemia.

Most of the experimental work that has been done with baboons has involved dietary manipulation of one sort or another. GRESHAM et al. [32] fed young baboons on various diets; some containing egg yolk, others having cholesterol in them. Cholesterol and phospholipid levels in the blood rose gradually for almost 6 months and then reached a plateau. The diet containing egg yolk produce the most spectacular lesions these consisted mainly of thoracic aortic fatty streaks that sometimes occupied as much as 25% of the surface (fig. 6). Not only were the blood lipids raised but the β-lipoproteins were also elevated. In the experimental animals the controls showing a preponderance of α- over β-lipoproteins.

Similar studies were made by STRONG et al. [33] who fed a number of diets that differed in the amounts of cholesterol and protein and in the nature of the added fat. The experiment was done with 40 young male baboons and lasted for 2 years. The higher cholesterol levels were found in those baboons

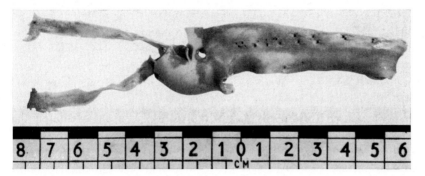

Fig. 6. Fatty streaking in the aorta of a baboon fed cholesterol and butter for 18 months. Sudan III/IV.

receiving the high-cholesterol, saturated fat and low protein diet. In general, they found that the most potent determinant of the level of plasma cholesterol was the dietary content of cholesterol. This is an important observation that reinforces the now widely accepted view derived from other species that dietary cholesterol is one of the principal factors in promoting the development of atherosclerosis in man. STRONG *et al.* also observed that the type of fatty acid in the blood and tissue was affected not only by the nature of the fat in the diet, whether it was unsaturated or not, but was also influenced by the level of cholesterol in the diet. For example the level of linolenic acid in these situations was lower in animals having high cholesterol diets. This could be regarded as another atherogenic factor by those who consider that the fate and the effect of cholesterol in tissues and in particular in blood vessels is determined by the type of fatty acid that esterifies it. In these experiments fatty streaks occurred in the aortas of the animals that had received high cholesterol diets. The fine structure of these was similar to those described for the rhesus monkey. They found fat-laden cells, smooth muscle cells and extracellular lipid together with collagen and fragments of elastic tissue. Also during some of the experiments they determined the long-term turnover rate, and the size of the readily miscible cholesterol pool together with the fraction of the miscible pool that was contributed by the diet. They did this by infusing 14C-labelled cholesterol into a vein and by feeding cholesterol that was labelled with tritium [34]. In general, they found that pool size and turnover rates were similar to those found in man. These early studies reinforced the view that the baboon is a good model for the study of the role of cholesterol in atheroma formation.

In 1966, MALINOW et al. [35] in common with many others in the field at that time were looking for a cheap, easy to maintain non-human primate for atherosclerosis research. They selected the squirrel monkey *(Saimiri sciureus)*. At once they appreciated the difficulty in obtaining a homogenous group for experimental purposes and emphasised the problems of determining the age of the monkeys, the effects of previous environmental conditions and so on. They also noted the difficulty of assessing accurately the amount of food consumed by the animals because of spillage, and described several diseases in these animals all of which may complicate the interpretation of dietary experiments. Finally, they reported the occurrence of atherosclerotic disease in their control animals fed on low cholesterol, low fat diet and whose serum cholesterol levels were well below 200 mg/100 ml.

However, they were able to show that diets rich in lipid did enhance the occurrence of naturally occurring disease even when no cholesterol was added to the diet. One important aspect of their experimental work was encompassed by the statement 'the findings in one artery were relatively independent of those in another'. The degree of aortic disease, for example, did not run parallel with the extent of involvement of the coronary arteries. This observation had already been made in man [36] and underlined the point that care should be taken to distinguish between aortic and coronary disease in research that is primarily designed to study the genesis of coronary atherosclerosis. The degree of aortic sudanophilia was increased if butter, margarine or cholesterol was added to the chow. If all three constituents were added then the effects on aortic sudanophilia were greater. Coronary artery sudanophilia, on the other hand, was increased only when cholesterol was included in the diet though again the effect could be reinforced by feeding butter as well. Feeding margarine with the cholesterol, if anything, protected the coronary arteries even though the blood cholesterol was high.

The degree of hypercholesterolaemia in the individual animals showed considerable variation, a fact previously reported in the rhesus monkey [37], nor was the extent of atherosclerosis in the experimental animals related to the levels of blood cholesterol.

In the year following the work of MALINOW et al. [35] the team from Winston Salem presented one of their many exhaustive studies of the experimental pathology of atherosclerosis in the squirrel monkey [38]. In particular, they clarified the effects of age and dietary protein on the production of experimental arterial disease. The animals were fed 0.5% cholesterol by weight and developed a significant rise in serum cholesterol levels. However, the young animals had a rise in cholesterol for the first 6 months of the experiment and during the next 6 months the levels declined. This contrasted with the older animals that had a steady rise over the 12-month experimental period. This suggested that the younger animals had a capacity to compensate for an increased dietary intake which may operate via an endogenous mechanism. When the dietary level of protein was high the effect of dietary cholesterol was to produce more aortic atherosclerosis than when the level of protein was low. If the diet contained no cholesterol the degree of aortic disease was not affected by varying the level of protein in the diet. This suggested some cholesterol-protein interaction as a factor in atherogenesis in this monkey. They observed the same phenomenon as that described by MALINOW namely that the addition of cholesterol particularly induced coronary atherosclerosis. Any sex differences in response that they noted were slight and not susceptible to statistical evaluation. Like MALINOW they found great variability within the experimental groups which they regarded as 'a basic difficulty in conducting nutritional studies on non-human primates'.

MARUFFO and PORTMAN [39] did work similar to that reported by the Winston Salem school and in particular confirmed the observation that the effect of feeding cholesterol to the squirrel monkey caused coronary artery lesions mainly in the distal rather than in the proximal parts of the vessels. This is an important point to be considered when attempting to extrapolate these results to the interpretation of human atherogenesis. They [39] studied the coronary arteries of squirrel monkeys which were fed either on an atherogenic or a non-atherogenic diet for 3, 6 and 8 months. Another series of animals was fed the atherogenic diet for 3 months which was then followed by periods of 3 or 5 months on a control diet before they were killed. The atherogenic diet produced accumulations of lipid both intra- and extracellularly in the coronary arteries. This was present in both intima and media but there was little associated cellular proliferation. The extent of the lesions correlated well with the terminal levels of plasma and aortic cholesterol unlike the findings of MALINOW et al. [35]. The vascular lesions were studied by electron microscopy and it is interesting that they were unable to find any smooth muscle cells in the experimentally induced lesions. A group of animals that were fed the atherogenic diet for 3 or 5 months and were then returned to the control diet did not show any regression of lesions. However, the lesions did not progress as did those fed the atherogenic regime for 8 months. It may well be that the time allowed for possible regression was too short in these experiments. In the regression group of animals small lesions were found in distal branches of the coronary arteries: they consisted of intimal accumulations of mucopolysaccharide associated with elastic tissue fragmentation and an increase of collagen fibres. These may represent an early lesion similar to the gelatinous lesions of man or alternatively they may once have been lipid-rich lesions that had subsequently regressed by loss of lipid from them.

This résumé of the results of three principle groups of workers using the squirrel monkey illustrates the difficulties that arise in interpretation of results. It is difficult to know whether the differences that they observe under essentially similar dietary conditions can be explained by varying age of animals used, conditions of housing, composition of the basic laboratory chow, duration of captivity before the experiments began and so on.

Some attempts to resolve problems of this sort have been made by comparative studies using several species in the same experiment. PORTMAN and ANDRUS [40] used three species of New World monkey: the cebus *(C. albifrons),* wooly *(Lagothrix lagothricha)* and the squirrel *(S. sciureus).* Where fat provided 45% of dietary calories it was found that coconut oil raised the serum cholesterol in all three species and the addition of cholesterol further elevated the serum level. However, the extent of aortic sudanophilia observed after 6 months on the diet was very different in the three groups. Whereas the squirrel monkey had the most extensive lesions the cebus had very few and the wooly monkey was intermediate in degree of involvement despite the fact that serum cholesterol levels were comparable in all three. A similar difference in the occurrence of lesions in these three species has been observed in animals taken straight from the wild state.

C.albifrons is curious in one or two respects when used as a model for atherogenesis. Experiments in this animal have been done with atherogenic diets [41, 42] using a variety of food fats. WISSLER *et al.* [42] compared the effects of coconut oil, corn oil and butter in the diets of cebus monkeys. All three diets had 0.5% cholesterol added and were fed for a period of 45 weeks. Atherosclerosis and hypercholesterolaemia were most pronounced in animals fed coconut oil, the changes were least in these fed corn oil and intermediate in those given butter fat. BULLOCK [43] studied the effects of age on the response of the cebus monkey to diet and was able to show that adult animals were much more susceptible to the effects of dietary cholesterol than young ones. The most striking and unusual effect was seen in the occurrence of coronary atherosclerosis. In the adults it was extensive and associated with significant stenosis of the vessels whilst in the young cholesterol-fed animals the disease was of slight degree. The peculiar susceptibility of the coronary arteries of adult cholesterol-fed cebus monkeys is not easily explained. No difference could be found in the cholesterol intake or absorption in the two groups but isotopic studies suggest a speedier turnover of cholesterol in the young animals [44].

There are several reports of differing responses of various species of non-human primate to dietary experimentation and they have continued up to the present time. NEWMAN et al. [45[compared arterial lesions and serum lipid levels in spider and rhesus monkeys on an egg and butter diet. There was a conspicuous species difference in response: after 12 weeks on the diet the mean elevations of serum cholesterol were 285 mg/100 ml for the rhesus and 17 mg/100 ml for the spider monkey. Aortas and coronary arteries of the rhesus were extensively involved with sudanophilic lesions and furthermore the proportion of cholesteryl oleate was greater in rhesus than spider when the palmitate level was lower. Lesions in the spider monkey were few; histologically the lesions in both species closely resembled one another. Clearly the rhesus monkey is most susceptible to cholesterol feeding; the diets used in these experiments contained 0.4% of the sterol. This sort of level is not greatly different from that in the usual American diet and WISSLER et al. [46] were able to achieve cholesterol levels of the order of 483 mg/100 ml in rhesus monkeys fed an average American diet. The authors conclude with the hope that more comparative studies will be made on lipid metabolism in such experiments in order to elucidate the pathogenesis of atherosclerotic lesions.

The thorny problem of selecting the most suitable non-human primate for atherosclerosis research will never be resolved until all the relevant variables concerned with atherogenesis and the factors that affect these parameters are fully worked out. Efforts to do this are continuing at the present time. KOTZE et al. [47] tackled the problem of the analysis of sera from 100 wild baboons of the genus *Papio ursinus*. They determined the fatty acid composition of cholesteryl esters, phospholipids, triglycerides and lipoproteins. They also investigated the lipid content of the high and low density lipoproteins. Though in general their results were in accord with those of previous workers there were differences which may have been explained by the fact that the animals used in this study were freshly trapped. From the point of view of man they showed considerable similarities in lipid chemistry reinforcing the view that the baboon is a likely contender for pride of place in atherosclerosis research.

Another approach to the problems raised by the distinct species differences in the serum lipid response to cholesterol feeding was to study the effects of diet on hepatic and intestinal lipogenesis in monkeys. COREY and HAYES [48] studied these phenomena in *C. albifrons* and *apella, M. fascicularis* and *S.sciureus* and found that synthesis of cholesterol by the liver and of cholesterol and triglyceride by the gut was greater in the New World cebus than in the Old World cynomolgus monkeys. They also demonstrated a large inhibition of cholesterol synthesis by liver slices of squirrel monkeys when they had been fed butter fat and cholesterol. Clearly there are many variables to be studied before a logical selection of a suitable primate for studies on human atherogenesis can be made. More will be said of these matters in the next section.

It is appropriate to note, at the end of this section on dietary experiments that lesions can be produced by feeding diets containing no added cholesterol [49] and also that when minute amounts of cholesterol are fed lesions appear in the absence of hypercholesterolaemia [50]. In the former case KRITCHEVSKY *et al.* [49] studied baboons *(P.ursinus)* fed on a semi-synthetic diet for a year. The diet contained 40% carbohydrate and this differed in each of the four groups. The animals were given labelled mevalonic acid at the start of the experiment in order to study cholesterol biosynthesis and the various classes of plasma lipids were also determined. All animals in the test groups showed hyperlipidaemia and aortic sudanophilia. Of the carbohydrates used fructose seemed to have produced the most extensive degree of aortic involvement.

The experiments of ARMSTRONG *et al.* [50] are equally provoking in relation to the possible role of cholesterol in atherogenesis. The fed high fat diets contained very small amounts of cholesterol (up to 129 mg/kcal for 18 months). The animals that received cholesterol in the diet did not develop hypercholesterolaemia and yet they did show intimal thickenings with aortic sudanophilia and an increased level of aortic cholesterol. The important point was, however, that the cholesterol content of high density lipoproteins fell whilst that of the low density lipoproteins was increased. They concluded that the intimal changes that they had observed were more likely due to a subtle qualitative alteration in the plasma lipoproteins rather than to any change in blood cholesterol levels.

This important conclusion will be discussed further in the light of experiments indicating the possible toxic role of β-lipoproteins on vascular smooth muscle, an effect that might be of considerable importance in atherogenesis.

In this section we have largely considered non-human primates that are susceptible to diet-induced atherosclerosis. Perhaps a detailed study of lipid metabolism, in the most resistant of them might also be profitable; such an animal is the marmoset [51].

Cholesterol Metabolism

A central common factor in dietary-induced atherosclerosis in non-human primates is an elevation of the serum cholesterol and deposition of the sterol ester in the arterial wall. EGGEN et al. [52] include a table in their paper which summarises this point. Most aspects of cholesterol metabolism have been studied by workers using non-human primates in order to answer a number of questions concerning gut absorbtion, plasma transport, arterial permeability and sterol metabolism by the vessel wall. All of these questions are pertinent to the problem of accumulation of cholesterol in atherosclerotic lesions.

Absorption was studied by MANNING et al. [53] who fed male rhesus monkeys on diets containing 1 mg cholesterol/kcal. Cholesterol-1,2-^{3}H was added to the diet for 195 days and it was observed that during the period of the isotopic steady state about 83–92% of the serum cholesterol appeared to be derived from that in the diet the amount absorbed being of the order of 127–192 mg daily. The extent of the serum cholesterol rise following cholesterol feeding varies amongst the primates. Man [54] and the baboon show small rises despite high cholesterol intake; rhesus monkeys are much more responsive.

Transport of cholesterol in duct lymph and serum was studied by FRASER et al. [55] who fed cholesterol together with a variety of food fats to rhesus monkeys for a few days. They cannulated the thoracic duct and also sampled the blood in order to determine the various lipoprotein classes in lymph and serum. Serum in animals with high cholesterol showed that low density lipoproteins transported the increased amount of cholesterol whilst the amount on the high density lipoproteins fell in the hypercholesterolaemic animals. In the lymph most of the cholesterol was transported by chylomicra. The cholesterol content of lipoproteins in both lymph and serum did, however, vary according to the type of fat that was fed. After the ingestion of corn oil, cholesterol in lymph was present mainly as the ester; after feeding coconut oil it was mainly free cholesterol. The form in which the cholesterol reaches the circulation from the lymph may be supposed to affect its subsequent metabolism. Free cholesterol readily transfers from chylomicrons to other lipoproteins as well as to cells and tissues. After transfer, the lecithin cholesterol acyl transferase enzyme causes esterification in the smaller lipoproteins the esterifying fatty acid being acquired from lecithin. Cholesteryl ester, on the other hand, stays in chylomicra which are swiftly removed from the circulation by the liver. So it may be that the recently ingested cholesterol from coconut oil-fed animals may be metabolised more slowly than that from the corn oil fed group. It also means that cholesterol from coconut oil-fed monkeys might be more readily available to permeate the vessel wall.

Whilst it has generally been assumed that the induction of arterial lesions requires an elevation of plasma cholesterol the recent work of ARMSTRONG et al. [50] already quoted casts doubt of this view. They fed rhesus monkeys on diets containing fat and small amounts of cholesterol and compared the results with those of animals on a cholesterol-free diet. They showed that lesions could be produced without elevation of plasma cholesterol levels in animals where there was a shift of cholesterol from high density to low density lipoproteins. They concluded that intimal changes in response to a very low cholesterol intake are more likely due to subtle qualitative changes in lipoproteins without elevation of plasma cholesterol levels.

The post-prandial handling of dietary fat was also studied by RAO et al. [56] who fed ^{131}I-labelled triolein in olive oil and then studied the ^{131}I-triolein tolerance curves. Considerable species difference was observed. Man exhibited the highest levels of blood radioactivity, rats the lowest and macaques (M.radiata), rabbits and chicks occupied an intermediate position. These changes are affected by absorption rate from the gut, clearance of chylomicrons by liver and adipose tissue, recycling of labelled fat as low density lipoproteins and tissue oxidation. If the absorption rates of fat are supposed to be constant for several species of animals [57] studies of this sort might further illuminate the important question of arterial wall-lipid content.

Many theories of atherogenesis that are held nowadays envisage the passage of low density lipoproteins from plasma into the arterial wall [52] and we shall consider the role of turbulence and shear as adjuncts to the process. The notion is that low density lipoprotein permeates into the intima and inner media but, due to the vascular structure, and in particular to the distribution of vasa vasorum the lipoproteins are unable to pass further through the media and may indeed remain incarcerated in the inner vessel wall thus damaging cells and initiating or aggravating atherosclerosis. It may also be that another factor which restricts the movement of low density lipoproteins is its combination with mucopolysaccharide in the vessel wall [58].

The finding of fine extracellular droplets of perifibrous lipid in the aorta with increasing age, the lipid composition of which approaches that of plasma S_f 0–12 lipoproteins, also supports the view of progressive insudation into and trapping of lipoproteins in the vessel wall [59]. Other evidence is derived from histochemical [60], immunoelectrophoretic [61] and immunochemical techniques [62] which indirectly demonstrate the presence of lipoproteins. SCOTT and HURLEY [63] were able to show in comatose, moribund patients that when plasma low density lipoprotein, labelled in the peptide component with ^{131}I was injected, equilibration occurred most rapidly with liver and spleen, but that intima and inner media of aorta and coronary arteries showed a gradual increase over the 14- to 16-day period of study. Higher levels were reached in the proximal aorta than in the coronary arteries. It was interesting that the level of radioactivity in the inner vessel wall was unrelated to the degree of intimal thickening or to the presence of atherosclerosis. It does appear then that a slowly exchanging low density lipoprotein pool exists in the primate arterial intima no matter what the diseased state of the vessel may be.

So far as non-human primates are concerned the chimpanzee seems to be the most promising animal for the study of lipoprotein changes and atherogenesis [64]. Of all the non-human primates STARE et al. [65] found most aortic and cerebral arteriosclerosis in chimpanzees. In addition, these animals are about five times as sensitive to cholesterol feeding than are rabbits [66].

BLATON et al. [64] fed diets containing 2.5% cholesterol and 14% butter and produced an elevation of plasma β-lipoproteins which carried an increased load of unesterified cholesterol. In addition, the plasma levels of cholesterol and phospholipid were increased but there was no increase of triglycerides or very low density lipoproteins. The lipid and fatty acid changes in the low density lipoproteins closely resembled those found in man so that the lipid and lipoprotein alterations that had been produced were a replica of the human situation.

The low density lipoproteins of the experimental animals carried more lipid per unit of protein than the control which led to a decrease of the mean lipoprotein hydration density. This effect may be important in reducing the mobility of these lipoproteins in the vessel wall and thus favouring cholesterol deposition and atherosclerosis.

It is unfortunate that the scarcity and cost of chimpanzees makes their use for atherosclerosis research so limited.

DAYTON and HASHIMOTO [67], in a critical review, consider much of the evidence about the entry of lipoproteins into the vessel wall. First of all they discuss the evidence that there is a flux of cholesterol from plasma to artery and they consider the discrepant proportion of free cholesterol and ester cholesterol in plasma and in the arterial wall which suggests that the flux is not due to the entry of lipoproteins but rather to a physiochemical change involving plasma lipoprotein and cholesterol in vascular endothelium. Previous work which was cited in this chapter purporting to demonstrate lipoprotein in the arterial intima may merely indicate a disassociated form of lipoprotein such as a peptide. DAYTON and HASHIMOTO [67] conclude '... lipoprotein peptide is know to enter arterial tissue, and cholesterol and its esters are known to enter. But it is not clear that any of the lipoproteins enter as the intact package'. Reviewing the morphological evidence up to that time (1970) they were unable to find any clear statement of the finding of lipid materials in pinocytotic vesicles of endothelial cells. However, they concede that in pathological states, where endothelial cell junctions may open, lipid may find its way into the intima. As for example the finding of lipid in the intima after various sorts of injury which had been preceded by the infusion of egg yolk [68].

Whether or not lipid gets into the intima from plasma it is difficult to assess the significance of it in the arterial wall. PORTMAN and ALEXANDER [69] demonstrated that between the time of birth and the adult state the lipid composition of the inner layers of the vessels of rhesus monkey changes in several respects whereas lipids from other tissues remained more constant with increasing age. Even though no macroscopic or microscopic lesions can be seen in monkeys on normal diets, nevertheless the lipid changes that occurred were similar to those found in the atherosclerotic state. These changes were an increase in free and esterified cholesterol increase in sphingomyelin and an increase in phosphatidylserine. In man also, SMITH et al. [70] found a better correlation between lipid accumulation in arterial intima with age than with the presence of intimal thickening. So it appears that lipid may accumulate in the primate intima without any visible evidence of disease and one is left wondering about the role of lipid accumulation in atherogenesis.

PORTMAN [71] has also asked this important question: 'Is it possible to make reasonable conjectures from these observations about the mechanisms of change in lipid composition and atheroma formation?' He summarises his view with a diagram and proposes two phases in lipid accumulation a pre-sudanophilic phase characterised by increases of free cholesterol and sphingomyelin and a subsequent sudanophilic phase in which there is an increase of phosphatidylcholine and cholesterol esters.

The first hypothesis, that sphingomyelin and free cholesterol increase together is based on several bits of evidence: the two sterols increase together in the earlier stages of lipid accumulation in the aorta and both are prominent in the plasmalemma and in its vesicles. It remains to be proved that increased plasma levels of the sterols increase the degree of pinocytotic activity by cells but it is likely. If cholesterol is added to phospholipid monolayers the film contracts [72]: perhaps free cholesterol causes buckling of the plasmalemma with the formation of pinocytotic invaginations. PORTMAN explained the accumulation of cholesterol ester in the second sudanophilic phase by the release of ester from lipoproteins that had entered the vessel wall and proposed that these esters would accumulate as the level of the tissue ester hydrolases was exceeded.

This hypothesis raises the problem of metabolism of cholesterol and its esters within the vessel wall. First of all there seems little doubt that cholesterol is synthesised by the arterial wall and that the amount is proportional to the cell mass suggesting that the synthesis is related to cell reproduction. Synthesis from precursors such as acetate and mevalonate has been reported [73]. Not only does the artery synthesise cholesterol but it also esterifies it and in severely atherosclerotic pigeon aortas, for example, this was the predominant process [74]. With isotopic studies it was found that the principal incorporation of label was into fatty acid esterifying cholesterol suggesting that acylation and transacylation was an important process in atheroma and ABDULLA et al. [75] have reported augmentation of LCAT activity during atherogenesis. It must, however, be borne in mind that these changes are very dependent on the kind of animal that is used. RAO and RAO [76] studied the total lipids and *in vitro* incorporation of acetate-1-^{14}C into different lipid portions of thoracic and abdominal aortas of monkeys *(M.radiata)*, rabbit, rat and chick. In all animals the percentage of total lipids in the thoracic aorta was higher than in the abdominal part but the monkey aorta had the lowest level. In all species the highest incorporation was into phospholipids and lowest in the cholesterol fraction; in general the monkey aorta incorporated more into phospholipids than other species. Results like these are difficult to reconcile with tendencies to develop spontaneous atherosclerotic disease in the animals that have been studied.

In general, however, it is well accepted that cholesteryl oleate is the principal ester that accumulates in atheroma. This is partly because of increased synthesis in the plaque [67] and partly to preferentially slow hydrolysis of cholesteryl oleate compared with other esters such as cholesteryl linoleate [77].

A number of investigators have demonstrated hydrolysis of cholesterol esters by the arterial wall which is, as BOWYER et al. [77] have shown, one important aspect of cholesterol metabolism that might be related to the accumulation of particular esters in atheroma. A depression of hydrolytic activity which was described following cholesterol feeding is another way in which removal of cholesterol ester from vessels can be hindered [78]. It does not appear that there is any mechanism for the degradation of cholesterol itself in the arterial wall. Despite this fact cholesterol can be induced to come out of early atheromatous lesions and it may be ejected through the intimal surface or less likely via the outer media and adventitia in lymphatics.

The situation concerning cholesterol metabolism in the arterial wall is one that will doubtless be clarified by further work in non-human primates. Present knowledge is, however, still compatible with the view that high circulating levels of cholesterol accelerate its entry into the arterial intima.

One of the most valuable techniques for the study of cholesterol metabolism which has been evolved in recent years is that of arterial perfusion. The technique was elaborated in 1965 by LOFLAND and CLARKSON [79] and independently in 1968 by BOWYER et al. [77]. LOFLAND et al. [80] studied perfused isolated segments of aorta and coronary arteries from *C. albifrons* fed on a diet including 0.5% cholesterol. They demonstrated that the vessels were able to synthesise fatty acids from acetate-1-^{14}C the rate being greatest in the vessels more susceptible to atherosclerosis namely the coronary arteries. Using mevalonic acid-2-^{14}C they established that aorta and carotid arteries could synthesise squalene. The aortas of cebus monkeys were shown to have cholesterol and its esters in increased amounts as atherosclerosis became more extensive. The free cholesterol form predominated over the ester in the ratio of about 3:1 which is the reverse situation in the plasma of these animals. When ^{14}C-cholesterol is given to such animals intravenously the isotope is found in a free:ester ratio of 3:1 after 37 days of experiment. This result suggested a propensity of the atherosclerotic vessel to admit and to retain certain forms of cholesterol.

Comparatively little work has been done with vascular perfusion techniques to study lipid synthesis by primate vessels and there is much to be learned not only of the types of lipid synthesised but also rate of synthesis and variations in synthesis along the course of the vessel. Perfusion of aortas of squirrel monkeys fed cholesterol with (1-^{14}C)-acetate and (2-^{14}C)-mevalonate [81] showed that the fatty acid composition of phospholipids and triglyceride was similar in both control and experimental subjects; the principal difference was the increase of cholesteryl oleate in the aortas of the animals fed cholesterol. The synthesis of ester was about six times that of controls. A most significant point was that these changes were demonstrable in arteries from experimental animals that were only slightly more diseased than controls.

Another recently developed method involving *in vitro* incubation studies of inactivated and reactivated enzymes has cast a little light on the rates of esterification of cholesterol [82]. Using this technique LACKO et al. [83] studied initial rates of cholesterol esterification in a variety of species. In general, they found, that the rate of esterification tended to increase with the level of serum cholesterol in all species. LCAT was particularly studied as it is a key enzyme leading to the production of cholesterol esters of plasma. Before this work appeared there had been several reports of plasma LCAT activity in various animals but there was little information on the initial rate of plasma cholesterol esterification. The present method which allowed equilibration of labelled cholesterol with the subjects own lipoproteins and then employed them as substrates answers this question. The rat has the highest fractional rate whilst the rabbit has much less. Man and pig have low rates. This relates to the tendency to develop atherosclerosis in these animals.

The role of the LCAT in the arterial wall as opposed to plasma is not yet fully understood. The enzyme may transport cholesterol from peripheral tissues to the liver or it may be concerned in the surface-volume ratio of very low density lipoproteins so that it may be involved in mechanisms that enhance or impeded atherogenesis. The finding of high levels of plasma LCAT in animals on atherogenic ducts may merely be a feed-back response caused by the elevated plasma cholesterol levels rather than any supposed atherogenic effect of LCAT itself. It may well be that the synthesis of β-lipoproteins and consequently plasma cholesterol and LCAT may be controlled by a common mechanism [84].

A variety of variables have in the past been related to atherogenesis, for example, arachidonic content of cholesterol esters [76], levels of aortic lysosomal enzymes [85] and the capacity for cholesterol absorption [86]. The role of LCAT seems to be protective in that the fractional esterification rates are higher in the species most resistant to the disease, namely the rat. Rabbits are a curious exception to this generalisation but this animal makes its cholesterol esters mainly in the liver and also LCAT levels in this species do not rise parallel to the increase of plasma cholesterol following an atherogenic duct.

STOKKE [87] studied LCAT activity in eight different species of animals. Unlike the previous workers he found the highest activity in plasma LCAT of monkey and man, however, he also examined liver acyl-CoA cholesterol acyl transferase and liver lysosomal acid cholesterol esterase and found no evidence of acyl-CoA:cholesterol acyl transferase in human liver. STOKKE concluded that at present it is difficult to relate the result that he found to the tendency of various species to develop atherosclerosis and emphasises that the different results obtained by workers with this method must be examined in the light of the preparation, amongst other things, of the heat-inactivated substrate source for measuring the enzyme. A similar situation exists with regard to the enzyme acyl-CoA:cholesteryl acyl transferase which is thought to be the dominant cholesterol esterifying enzyme in the aortic wall of some animals. HASHIMOTO and DAYTON [88] studied the enzyme aortic activities of aortic microsomes isolated from a variety of species, some susceptible to atherosclerosis and some not. They found that the microsomal cholesterol-esterifying activity was highest in the species most resistant to atherosclerosis namely the rat and concluded that such activity, far from predisposing to atherosclerosis, may have a protective effect. They suggested a new hypothesis on the basis of their results namely that free cholesterol was itself atherogenic and that esterification could be regarded as an adaptive change.

There has been much discussion in times past about the ratio of phospholipid to cholesterol as a factor in atherogenesis. A high phospholipid/cholesterol ratio being advantageous in that it is postulated that the phospholipid solubilizes cholesterol in the vessel wall [89]. PARKER et al. [90] studied the metabolic activity of myointimal cells from atherosclerotic lesions and showed the incorporation of linoleic acid-1-^{14}C into phospholipids in both normal and atherosclerotic vessels it being greatly increased in atherosclerosis. Using labelled glucose they also showed an increased uptake by the diseased vessels which paralleled an increase of oxygen consumption. This increased metabolism and synthesis of phospholipid was associated with the appearance of myelin figures in myointimal cells as shown by electron microscopy. It could be argued that this was a compensatory mechanism for solubilizing cholesterol in the vessel wall and NEWMAN and ZILVERSMIT [91] support this view by contending that the phospholipid was not being used for the synthesis of cell membranes because these were only minor changes in DNA and protein content in the atherosclerotic vessel. BORENSZTAJN et al. [92] made a study of the incorporation of ^{3}H-thymidine into DNA in rhesus monkey aortas in which atherosclerosis had been induced experimentally. There was an intense proliferation of smooth muscle cells in these lesions and the proliferation occurred as early as 15 days after the start of the feeding of the experimental diet which contained peanut or corn oil. However, no differences in aortic DNA-specific activity could be recorded even though there was an increased incorporation of ^{32}P into phospholipid. It was interesting that the incorporation of ^{32}P into phospholipid and RNA was less in the abdominal aorta than in the thoracic. This may be a reflection of disordered metabolism and decreased phospholipid synthesis of this part of the vessel which could be predisposing factors for atherosclerosis in the distal aorta.

The significance of phospholipid synthesis in relation to atherogenesis is not clear and discrepancies with the proposed protective role of phospholipid exist in non-human primate experiments. For example PEETERS et al. [93] using baboons fed atherogenic diets observed a rise in cholesterol/phospholipid ratio in α-lipoproteins and the ratio fell in the β-fraction. This is contrary to the view that a rise of cholesterol/phospholipid ratio is a prerequisite for atherogenicity.

This review of cholesterol metabolism indicates a number of differences in some parts of the process in non-human primates and in other experimental animals. Much is still to be learned about the role of cholesterol in atherogenesis in the primate. However, any species that is used is likely to produce meaningful results provided that a careful comprehensive study based upon sound methology is used.

To date little has been done on the actual physical process of permeation of substances through the endothelial surface. We have already referred to increased pinocytotic activity in cells of the atherosclerotic vessel wall. It can be shown that an atherogenic diet will lead to percolation of intravenously injected collidal iron into the endothelial cells [94]. Furthermore, it has been shown repeatedly that intravenously injected Evans blue which binds to serum albumen permeates the endothelial layer at those sites where atherosclerotic lesions tend to occur namely at branches and divisions in the arterial tree. SOMER et al. [95] have done a good deal of work on this matter and have, in addition, demonstrated that at sites where dye uptake is greatest there is also an increased incorporation of ^{3}H-cholesterol into the vessel wall.

Only occasional references have been made in this book to the possible role of mucopolysaccharides in atherogenesis. Recently it has been suggested that these substances within endothelial cells may play a part in transcellular transport. Hyaluronic acid and chondroitin sulphates have recently been shown in endothelial cells [96] and it has been demonstrated that increased hyaluronidase activity in the blood has an atherosclerosis promoting effect [97, 98]. It is well known that acidic mucopolysaccharides restrict the movement of molecules in the ground substance of connective tissues. The question is whether these substances restrict molecular transport in endothelial cells. KLYNSTRA [99] tried the effect of hyaluronidase on the pig aorta and then incubated the vessel filled with a diluted india ink. He then determined the distribution of the ink and the amount of mucopolysaccharide in the various parts of the vessel wall. The amount of mucopolysaccharide was reduced in areas where the dye had permeated. The mode of action of these substances is debatable; they may form a glycocalyx on the cell [100] or may be part of the pore-like structures of the cell membrane [101].

The reacting surface of the vessel wall is the endothelium. More work on variations in cell turnover, metabolism and cell layer integrity in the non-human primate cannot fail to help our understanding of atherogenesis.

Haemodynamic Studies

So far we have considered the reacting tissue namely the arterial wall in the pathogenesis of atherosclerosis in non-human primates. It is surprising how little has been done on the role of haemodynamic factors and blood coagulation in atherogenesis in the primate order particularly as hypertension is considered to be a potent risk factor in the genesis of atherosclerotic disease in man. There are few reports on the production of experimental hypertension in monkeys and yet GOLDBLATT [102] as early as 1937 had started this sort of work in the rhesus monkey. He adapted his method of clamping the renal artery of dogs, for which he is well known, to the rhesus monkey, and was able to produce sustained elevation of the systolic and diastolic blood pressures. One of the problems that he encountered was the small size of the renal artery of the macaque so that it was difficult to know whether the vessel had been constricted or totally occluded by his clamps. Further work appeared in 1961 when McGILL et al. [103] observed lipid-containing intimal lesions in the aortas and large blood vessels of hypertensive rhesus monkeys. They used adult animals and performed sequential bilateral renal artery constriction. The animals survived for 19–49 months and 10 of the 14 animals became hypertensive. The lesions that they saw were very much like fatty streaks or spots and were sudanophilic. In addition, fibrous or pearly plaques were also seen and they were most prevalent in the abdominal part of the aorta. Histological examination of the fatty streaks showed features that we have already discussed, namely smooth muscle cells, and mucopolysaccharide material. Some cells contained lipid, early collagen formation was found and the internal elastic lamella was intact. The more fibrous lesions contained smooth muscle cells and more mucopolysaccharide than the fatty streaks; the fat content of the lesions was variable.

Some of the animals were given repeated injections of hog and dog renin but this did not aggravate the arterial lesions despite the fact that such injections had been reported to elicit arteriolar lesions [104]. The amount of cholesterol in the diet of these animals was small and the authors concluded that hypertension was a major factor in the production of increased atherosclerosis in the hypertensive monkey.

The principal difference between the lesions in hypertensive monkeys and those of man was the presence of abundant mucopolysaccharide and in the older lesions there was less lipid, collagen and elastic tissue than in man. These differences probably are mainly a reflection of the speed of development of lesions rather than representing a fundamental difference from the human disease.

Pressure, turbulence and shear on the intima need further study in primates. Related to these factors is the role of platelets and blood clotting in initiating or augmenting atherosclerosis. Changes in pressure producing turbulent flow might be expected to promote collision of formed elements in the blood, in particular platelets, with the endothelial surface. There is no doubt that endothelial damage leads to platelet accumulation at the site of injury. PRATHRAP [105] made the first studies in the monkey *(M.irus)* on the effects of inserting a suture through the wall of the femoral artery. 66 animals were examined from 1 h to 2 years after the operation and he observed the gradual transition of platelet-rich thrombi at the site of injury into hard fibrous atherosclerotic plaques which did not contain lipid. These animals were normocholesterolaemic and when hypercholesterolaemia was produced by diet lipid accumulated in the lesions [106].

A great deal of experimental work has been done in times past on the possible thrombogenic origin of atherosclerotic plaques but very little in monkeys. In view of extensive recent studies on the structure, properties and chemistry of platelets the incrustation hypothesis probably needs modification [107]. As early as 1961 we suspected that platelets in addition to being incorporated as thrombi might of themselves damage the vessel wall. Platelets contain 5-hydroxtryptamine that is known to alter endothelial cell permeability and recently they have been implicated in the inflammatory reaction producing proteases that may also affect cellular integrity [108].

Hypoxic Injury

The view has been held for some years that experimentally induced tissue hypoxia may accelerate atherosclerosis in cholesterol-fed animals [109–111]. It has been held that hypoxia may directly induce the disease

[112]. If rabbits are fed cholesterol and also exposed to carbon monoxide they have higher levels of cholesterol in the serum and in the aorta than animals given cholesterol without CO exposure. In addition the hearts were heavier, there was more coronary atherosclerosis and more areas of myocardial necrosis were found [113]. WEBSTER et al. [114] exposed cholesterol-fed squirrel monkeys to intermittent dosage of carbon monoxide for 4 h a day, 5 days a week over a period of 7 months. The extent of coronary atherosclerosis was aggravated in the CO-exposed animals but the degree of aortic disease was unaffected. The hearts of the CO-exposed monkeys were heavier and there was more lipid in the coronary arteries and this had produced significant stenosis and electrocardiographic changes. These changes consisted of delayed ventricular depolarisation and right bundle branch block. The exacerbating effect of CO exposure on the coronary arteries was similar to that described by ASTRUP et al. [113] in the rabbit. However, there was no evidence of myocardial necrosis as ASTRUP had found. If the period of exposure of the squirrel monkeys had been longer, myocardial lesions might have been found but these can be difficult to assess, as we have already indicated, because focal areas of necrosis and cellular infiltration due to parasitic invasion not infrequently occur in the hearts of squirrel monkeys.

These studies do not explain how CO exposure operates to enhance coronary arterial disease. However, the effect on the coronary arteries was not so much as to increase the numbers but rather the size of the lesions due to lipid accumulation. Hypoxia is a well recognised cause of fatty change by interfering with the tricarboxylic acid cycle and there is evidence that CO exposure inhibits the production of lactic dehydrogenase [115] which is a catalyst for the production of ATP. This then may be one explanation. There is evidence also that atherosclerotic aortas have an increased oxygen consumption so that hypoxia may have relatively a more potent tissue-damaging effect in such diseased areas [116]. Another possible way in which hypoxia might exert an effect on the arterial wall is through lipoproteins. LAZZARINI-ROBERTSON [117] has shown recently by means of organ cultures of human atheromatous tissue that the uptake of lipoproteins is increased by the culture when the conditions are rendered hypoxic. We have already discussed the potentially toxic effect of β-lipoproteins for cells so that the process of lipoprotein uptake might operate to aggravate the atherosclerotic process.

The increased cardiac weights could not be related to an increased packed cell volume. Probably because of the relatively short exposure to CO the monkeys did not develop polycythaemia. Another surprising factor was the absence of ST elevation in the electrocardiogram which might have been expected with the cardiac hypertrophy and which is certainly seen in cebus monkeys developing cardiac hypertrophy [118].

There are very few reports of the role of hypoxia induced by carbon monoxide in atherogenesis in monkeys. It is, however, an important topic particularly in relationship to the known association of smoking and coronary artery disease in man. Levels of up to 20% carboxyhaemoglobinaemia can be achieved during bouts of smoking and work is now going on in several centres to study the effects of CO exposure on atherosclerosis in non-human primates. A recent publication from Copenhagen [119] concerns the effects of exposing *M.irus* to 250 ppm of CO for 2 weeks. The main findings in the coronary arteries were detected by electron microscopy and consisted of subendothelial oedema with separation of endothelial cells at their junctions. Monocyte-like cells were seen in the subendothelial space and some contained cytoplasmic droplets of lipid. No changes were detected at this early stage in elastic tissue nor in medial myocytes. It is interesting that the cells in the subendothelial space which had appeared following this brief injury had none of the characteristics of smooth muscle cells.

It may be that myointimal cells would appear at a later stage in the process. Longer term exposures are being at present done by BOWYER in Cambridge and may provide the answer to this question.

The Role of Dietary Carbohydrate

Excessive intake of carbohydrate tends to promote hypertriglyceridaemia which is sometimes associated with coronary artery disease in man. In very few of the dietary experiments that we have so far discussed was hypertriglyceridaemia a feature of the animals developing atheroma. Few experiments have been done on the effects of dietary carbohydrates on the arteries of non-human primates. Adult baboons fed for several months on a diet containing sucrose became very obese but the degree of aortic sudanophilia was slight [120]. The degree of obesity in these animals did not correlate well with the extent of atherosclerosis and this is often the case in man.

Nevertheless, elevated serum lipid levels and/or abnormal carbohydrate insulin mechanisms have been observed increasingly in the obese [121], and PETERS and HALES [122] also found that patients with coronary artery disease had an elevated level of plasma insulin both when fasting and 30 min after an oral dose of glucose. Relevant to this are the experiments of HAMILTON *et al.* [123] who produced obesity and hyperinsulinaemia in *M.mulatta*.

They used 19 male and 2 female monkeys which were arbitrally classified as obese if the body weight exceeded 15 kg. Two groups of obese animals were studied. The first had deliberate injury to the ventromedial part of the hypothalamus the second group became spontaneously obese at the age of 12–14 years. Two of the latter animals developed diabetes mellitus. Studies of the free fatty acids, of the spontaneously obese group showed that the levels were raised and both groups of obese animals developed hyperbeta and hyperprebeta lipoproteinaemia which was associated with elevations of serum cholesterol and triglyceride levels. All of these changes were considerable in the two diabetic animals that had not been treated with insulin.

HOWARD et al. [124] fed baboons on atherogenic and control diets and then did glucose tolerance tests by giving 2 g/kg glucose by stomach tube. Both groups of animals, experimental and controls, had high levels of plasma insulin in the fasting state and high values half an hour after glucose dosage. Even though the control and experimental groups showed no differences, the insulin response was much greater in the baboon than those observed in man and the rabbit. This makes it difficult to support the proposition that a high plasma insulin response is related to a tendency to develop atherosclerotic disease.

Experiments to do with carbohydrates and atherosclerosis are in general complicated by the fact that a number of factors have been studied at the same time. LEHNER et al. [125] investigated the effects of insulin deficiency, hypothyroidism and hypertension on atherosclerosis in the squirrel monkey in an attempt to clarify conflicting reports about the relationship of human diabetes mellitus and hypothyroidism to the occurrence of atherosclerosis. Few have used non-human primates for this sort of work which is surprising for as early as 1958 GILLMAN et al. [126] studied the relationship of hyperglycaemia to hyperlipidaemia and ketonaemia in depancreatised baboons. In LEHNER's experiments diabetes mellitus was induced by alloxan. They fed the animals 1 mg of cholesterol per calorie for over 3 years and found that hypothyroid and insulin-deficient monkeys had greater concentrations of serum cholesterol and β-lipoprotein than the controls. The hypertensive and control animals produced similar results. All three groups of experimental animals had more severe coronary and aortic disease than the controls, the disease being worse in the insulin-deficient animals. The histological structure of the aortic lesions differed from one group to the other and broadly speaking the results were similar to those seen in man. The hypertensive animals had musculoelastic lesions with less lipid than in the other two groups. Some of the insulin-deficient animals had typical atheromas with lipid gruel overlain by fibrous caps with adjacent calcification and medial thinning. As has been described before in the squirrel monkey the bulk of coronary lesions lay in the intramyocardial branches of the vascular tree and were histologically similar in all three experimental groups.

So far as lipid concentrations were concerned the highest levels of serum cholesterol were observed in the insulin-deficient animals and this is consistent with reports of diabetic men and animals such as rats, rabbits, and baboons [126]. The lipoprotein pattern of these animals showed an increased β-lipoprotein level but no prebeta band so that the situation that had been reproduced resembled type II hyperlipoproteinaemia in man. We have already referred to the effect of induced hypertension on the arteries of monkeys and this work confirms the results of McGILL et al. [103].

There is some debate about the relationship of hypothyroidism to severe atherosclerosis in man [127] though some of the most severely occluded coronary arteries that I have seen have been in patients with myxoedema where the lumen of the vessel is reduced to a slit the rest being occupied by lipid. In LEHNER's study [125] the most severe coronary disease was found in the hypothyroid squirrel monkeys.

The alloxan-diabetic animals had an abnormal glucose tolerance as might be expected. This was related to the serum cholesterol concentration; furthermore, the severity of atherosclerosis was related to the abnormality of glucose tolerance. The degree of atherosclerosis may merely be due to the elevation of plasma cholesterol but could equally be related to hyperglycaemia. In man EPSTEIN et al. [128] have reported elevations of blood glucose to be an independent risk factor in the development of coronary heart disease. They say it is at least as important as hypercholesterolaemia or hypertension.

LEHNER et al. [125] in their experiments have managed to keep the three groups of risk factors that they studied fairly well separated. Even so they ran into problems in the interpretation of their results because the hypothyroid animals became hypertensive as compared to the other two groups and the insulin-deficient animals also developed mild hypertension as compared to controls.

It is always difficult when planning dietary experiments to resist the tendency to introduce a large number of variables. It is particularly important when devising experiments on non-human primates if only because of the difficulty in setting up large enough groups to produce viable statistical results. Multiple factors in an experiment may create problems if one augments or diminishes the effects of the other. A recent study of the comparative lipid response of four primate species of young monkeys (squirrel, cynomolgus, cebus and Spider) to dietary changes in fat and carbohydrate though beautifully done and well documented illustrates the problem of this kind of work [129]. COREY et al. [129] attempted to reduce as many variables as possible by defining the age and origin of their monkeys. They then fed low fat and relatively low cholesterol diets; in addition, the amount and type of carbohydrate was varied. The results of the experiment confirmed previous reports already discussed of the considerable interspecies variation that exists among primates in the serum cholesterol response to types of dietary fat and cholesterol supplements. The relative resistance of the spider monkey [130] was also confirmed. Neither saturated fat nor cholesterol were remarkably hypercholesterolaemic for this animal until sucrose was substituted in the diet as a source of carbohydrate. The difficulty in deciding the significance of this observation is complicated by the knowledge that progressive elevation of cholesterol levels occurs in cebus [131] and squirrel monkeys [132] when fed saturated fat without sucrose. This has been thought to be associated with progressive maturation of the animals during the course of the experiment.

Summarising then, the additive effect of dietary cholesterol was not observed in spider monkeys, was only slight in the cebus, was moderate in the squirrel monkey and considerable in the cynomolgus.

So far as triglycerides were concerned, feeding a high carbohydrate diet did not induce hypertriglyceridaemia in these animals, unlike the situation in man. Sucrose did, however, potentiate hypercholesterolaemia. Other workers, however, have observed hypertriglyceridaemia in monkeys following sucrose feeding and also elevations of plasma phospholipids [133]. Clearly it is a complex problem to try to unravel the differences in results of various workers. However, explanations for such discrepancies may provide useful information about the role, if any, of dietary carbohydrate in atherogenesis.

The Effects of Vitamins and other Substances

From the earliest days of the use of non-human primates until the present time views have been held about the possible atherogenic effects of vitamin deficiency or excess. In 1956, RINEHART and GREENBERG [134] summarised their work on the effects of feeding pyridoxine-deficient diets to 40 rhesus monkeys. They described intimal changes in large and small arteries consisting initially of the appearance of intimal mucinous material followed by proliferation of cells and connective tissue fibres, both elastin and collagen. Occasionally the elastica of the vessel wall was degenerate; there was a little lipid and slight calcification. They postulated that the mucoid material might favour lipid deposition which is a view that we have already examined. In relation to the preceding section of this chapter it is interesting to note that the diets used by them contained 7.3% sucrose.

RINEHART and GREENBERG's work on pyridoxine deficiency was not sustained by the experiments of MANN et al. [135]. The pyridoxine (B_6) hypothesis was attractive because it is known that the vitamin plays a role in the formation of the cross-linkages of elastin. It is needed as a cofactor for the enzyme amine oxidase which is essential for their formation. GREENBERG et al. [136] tried to determine a decreased synthesis of elastic tissue in the aortas of B_6-deficient young adult rhesus monkeys by measuring desmosine and isodesmosine levels. They found no evidence of a lowered concentration of these substances in their experiments. This may be because the turnover of elastin in the young adult was virtually at a standstill. MANN et al. [135] formulated another deficiency hypothesis namely that lack of choline in the diet was an atherogenic factor [137]. They fed diets high in cholesterol and low in sulphur-containing amino acids to *Cebus fatuella*. The animals developed lesions and hypercholesterolaemia. The lesions were mainly in the aorta and consisted largely of fat-laden cells filled with lipid; the connective tissue response was scanty. Subsequent work did not confirm the effects of choline deficiency. Indeed, in rats fed high fat, low choline diets lesions were produced which bear little relation to atherosclerosis consisting largely of medial calcifications [138]. These lesions most resembled those seen with hypervitaminosis D.

The extensive literature on experimental atherosclerosis in animals contains frequent accounts of lesions that have wrongly been called atherosclerotic. These calcifications that we have already mentioned are primarily medial in origin consisting of fine deposits of calcium on straightened, fractured, inner elastic lamellae with variable adjacent accumulations of mucopolysaccharide. Such lesions have been described in pigs, aged female rats, rabbits, cattle and in monkeys [118]. Although they may form a basis for the subsequent intimal accumulation of lipid if hypercholesterolaemia is induced it is unlikely that the medial lesion is a primary atherosclerotic event.

An excessive intake of vitamin D_3 in the diet will induce such medial calcifications and it is important to bear this in mind when devising diets for atherogenetic research. It seems that non-human primates vary in their need for and sensitivity to such vitamins. For example squirrel monkeys given large doses of D_3 developed little calcification apart from occasional nephrocalcinosis whereas in the cebus monkey extensive calcification of the heart and aorta and other tissues occurred [139]. Vitamin D_3 also has a 'lipid-loading' effect [140], which might cause considerable difficulty in interpretation of dietary experiments concerned with hyperlipidaemia.

The reported effects of vitamin deficiency and excess on arterial disease in various species is fully summarised by KIRK [141] in his monograph in this present series. He presents a table which clearly illustrates the confused state of modern knowledge about this matter.

Deficiency and excesses of various materials such as copper, magnesium, trace elements and protein deficiency have variously been reported to augment atherosclerotic diseases in animals but little has been done so far in the non-human primate [142].

Differences in the tendencies of men in hard and soft water areas to develop cardiovascular disease have been reported [143]. Here again there is some confusion in the literature but the main points seem to be that soft water consumption predisposes to myocardial infarction and cardiac dysrhythmias. How this effect is mediated is unclear but the probability exists that an excess or lack of trace elements such as cadmium and other metals might be operating here. We attempted to determine differences in the degree of arterial disease in cebus monkeys given Cambridge hard water and soft water brought from Stirling in Scotland. We were unable to demonstrate any differences.

Several exogenous chemical substances have effects on vessel walls and it is most appropriate to discuss them here. Perhaps that most often tried is β-amino-propionitrile derived from the seeds of *L. odoratus*. Very little work has been done with this agent in the non-human primate. The effect of the substance is to produce a generalised connective tissue defect by inhibiting cross-linking in collagen and elastin, rather similar to the effects of vitamin B_6 deficiency. These effects have been well documented; more controversial is the effect on tissue mucopolysaccharide some reporting an increase others describing a decrease of these mucosubstances. The confusion is probably due to the variety of tissues studied and to the varied ways of administering the lathyrogen. BENTLEY *et al.* [144] review the situation and conclude that because the effect of the lathyrogen is by inhibition of amino-oxidase which causes oxidative deamination of amino groups of lysine the mucopolysaccharide changes are secondary. The effect of the enzyme inhibition is to reduce the number of aldehyde groups available for cross-linking. When this occurs in the aorta the collagen is weak and the animal eventually dies of aortic rupture. Accumulation of mucopolysaccharide reflects attempts at repair in the damaged vessel wall.

LALICH et al. [145] produced intimal disease in the abdominal aorta and common iliac arteries of infant rhesus monkeys by feeding β-amino-propionitrile. They showed oedema of the intima with destruction of inner elastic tissue and haemorrhage with subsequent haemosiderosis in the damaged areas. In their experiments the abdominal aorta and iliac arteries were more susceptible than the thoracic aorta whereas in the rats fed with the substance the aortic arch is more often affected. The changes that they observed were roughly the same in animals fed for 6 or 12 months. Presumably this is due to the fact that the lathyrogen acted at the most formative period of the growing aorta namely in the early months.

The β-amino-propionitrile model is useful for the subsequent study of hyperlipidaemia or some other factor on injured arteries but it is debatable if it can be considered primarily as a model of atherosclerosis. Certainly it is not a morphological replica, but there may be more subtle biochemical similarities. It has been suggested, for example, that the smooth muscle cell can synthesise elastin [146] and from a study of enzymes in the vessel wall of animals given β-amino-propionitrile WICKS and GARDNER [147] conclude that the effects produced might be the result of altered lysosomal permeability of medial cells. PETERS and DE DUVE [148] used cholesterol-induced atherosclerosis as their model and suggested that pinocytosis of low density lipoproteins by smooth muscle cells may be an early event in atherosclerosis. This lipid material collects in smooth muscle cells in their model because of a defect in lysosomal esterase.

We have already proposed that biochemical models might be much more productive than purely morphological studies though clearly the two should be complementary. It is interesting how two seemingly remote agents such as cholesterol, on the one hand, and β-amino-propionitrile on the other come together in a final common path of action.

Various Sorts of Vascular Injury

To some extent we have discussed this matter in the previous section but it is important to separate the issue of vascular injury in order to emphasise the point that many sorts of injury can result in a final process which is the atherosclerotic plaque. CONSTANTINIDES [142] lists the wide range of injury of one sort or another that has been used and emphasises the problem of trying to decide which vascular injuries if any are responsible for the disease in man. He concludes: 'One can compare an injured artery with a red swollen finger; it is impossible to tell by just looking at that finger whether or not the inflammation was caused by the bite of a scorpion, by radiation, by a crushing injury, by an infection or by hundreds of other factors.' In other words, the atherosclerotic lesion can be regarded as the basic response of an artery to many forms of damage.

An interesting toxic agent which damages vessels and has been used in many animals is derived from the seed of *Crotalaria spectabilis* which is a legume used as a nitrogen-fixing plant to improve the quality of soil. It is toxic to many species but had not been tried in non-human primates until 1965 [149]. At a concentration of 0.25–1.0% in the diet the substance acts as an hepatic toxin for rhesus monkeys causing depression of albumen synthesis, inter alia, and generalised oedema. However, the substance has also been observed to have potent endothelial toxic effects in several species but this was not observed in the present study. Later however, the same workers [150] by using the pyrrizolidine alkaloid monocrotaline from Crotalaria given intravenously were able to produce occlusion of hepatic veins throughout the liver; both large and small veins were affected. Early changes were of damage to endothelium; late changes were of fibrous intimal thickening leading to vascular occlusion. To date there are no reports of the effects of these alkaloids on the arteries of non-human primates, but they might be considered as useful tools for localised endothelial injury in temporarily isolated arterial segments.

For many years it has been well known that hypersensitivity might lead to the production of arterial disease and presumably because the clinical effects were mainly those related to small vessel damage these latter received most attention. Little has been said until recently about the possibility of immunological injury as a factor in atherosclerosis. Nowadays, for some, it is considered to be the most potent precursor of the disease.

In serum sickness lesions can appear in arteries in as little as 3 days; they consist of endothelial proliferation, occasional necrosis and leucocytic infiltration [151]. Accumulation of antigen-antibody complexes in the vessel wall was thought to be responsible. LEVY [152] and subsequently others looked at the possibility that a serum-sickness type of response might potentiate atherosclerosis in the rabbit. Such studies comprised feeding a cholesterol-enriched diet accompanied by multiple injections of a foreign protein. Cholesterol alone produced lesions in 4 weeks; when reinforced by injections of protein lesions appeared in 2 weeks [153]. So one stimulus clearly potentiated the other. The proposition is then that the effects of the hypersensitivity is to produce an inflammatory response in the inner vessel wall followed by a proliferative repair process that ultimately forms the mature atherosclerotic plaque. All who have studied coronary artery disease in man will from time to time have noted acute or subacute inflammatory changes in the lesions or in the adjacent adventitia [154]. It is also true that a variety of inflammatory mediators such as histamine and 5-hydroxytryptamine will increase endothelial permeability probably by opening intercellular junctions. This allows plasmatic materials to enter the intima and to produce the early atherosclerotic phase.

It does then appear that the atherosclerotic lesion can be regarded, in its early phases, as an inflammatory response. We have already suggested that immune complexes might be one mode of intimal injury that sparks off the process. HOWARD et al. [155] produced some evidence for this hypothesis in the baboon. A hypercholesterolaemic diet alone produced scanty aortic sudanophilia after a 2-year experiment. The process was enhanced by repeated injections of bovine serum albumen intravenously. In this case it was postulated that deposition of immune complexes in the vessel wall caused the atherosclerotic injury.

The entry of damaging immune complexes into the intima is aided by inflammatory mediators such as 5-hydroxytryptamine and a source for this could be blood platelets. Immune complexes themselves cause 5-hydroxytryptamine release from platelets and this process is aided by complement factors. At the beginning of this monograph we suggested that blood platelets might cause the single injury to the vessel wall that triggers of the atherosclerotic process. This hypothesis has been proposed to several generations of Cambridge medical students over the years and it is gratifying to see the evidence for it accumulating.

We are still left with the question of the initiation of the platelet 5-hydroxytryptamine release reaction. POSTON and DAVIES [156] hold the view that the trigger is immunological produced by allergy to absorbed unaltered milk proteins or perhaps to microbial agents or to tobacco leaf protein.

It seems more likely, however, that any immunological mechanism involved will be lymphocytic T cell mediated rather than B cell mediated. In favour of this is the extraordinary acceleration of the development or coronary atherosclerosis in the transplanted heart of a young Cape-coloured man into a middle-aged man with severe myocardial disease. Over the space of 19 months severe disease developed in this heart and it was reasonable to believe that the coronary arteries were near normal before the heart was transplanted [157]. Also a constant feature of immunologically rejected transplanted kidneys is the presence of severe obliterative intimal thickenings [158].

An extraordinary twist to the story of arterial injury was given by BENDITT and BENDITT [159]. They showed in four patients that the smooth muscle cells of atherosclerotic plaques were monoclonal in origin. This they did on the basis of the production of a single enzyme type by the cells which were not present in the cells of the adjacent normal vessel wall. They rule out the possible effects of mosaicism by taking samples of a size larger than known mosaic patches in the aorta. They therefore propose that the plaque cannot be an inflammatory response to injury because these cells would be expected to be polyclonal in origin. The conclusion is that the atherosclerotic lesion is a result of local mutation perhaps by the effect of viruses or chemical mutagens. In support of their view is the occurrence of lesions at specific sites in the vascular tree where endothelial turnover and hence the chance of mutation are increased. The augmenting effects of elevation of blood lipids and hypertension can be similarly explained on the basis of increased endothelial turnover and hence mutation.

The potentiating effect of arterial injury on lipid deposition has been demonstrated in a variety of experimental models but not often in the non-human primate. Cox et al. [160] showed that in the presence of a moderate hypercholesterolaemia in the rhesus monkey lesions developed in 3 months but if segments of artery were injured by freezing lesions developed in these in as short a time as 3 weeks.

The Effects of Hormones and Stress

Hormone release and stress are often related so that we shall consider both together in this section. In both cases the field is somewhat confused because of the variable effects of hormones on atherogenesis in different animal species and the lack of any precise definition of stress and of the appropriate variables that should be used to assess it.

The clearly defined middle-aged human male preponderance of atherosclerosis over the female is rarely observed in other primates. This is often, no doubt, because much of the work that we have already discussed dealt with juvenile or young adults and also because the degree of disease was slight. However, even in TAYLOR's classical experiments with rhesus monkeys, where severe lesions were produced there were no observable sex differences. Indeed, the only animal that developed myocardial infarction was female. Of the many reviews of non-human primate atherosclerosis few refer to sex differences in the occurrence of the disease and none record any strikingly significant effect of sex. Furthermore, little has been done to study the effects of changing levels of sex hormones on arterial disease in these animals.

STAMLER [161] reviewed the work done with oestrogens in cockerels showing a diminition of coronary atherosclerosis in animals fed cholesterol and given oestrogen and emphasised the difference in response of aorta and coronary arteries in that animal. Much has been done in many other animals using thyroidectomy, ^{131}I dosage, administration of catecholamines, cortisone and thyroxine, but apart from the effects of thyroid ablation, which enhances atherosclerosis, the non-human primate does not figure largely in these reports.

Because of the known effect of adrenaline in raising plasma free fatty acid levels experiments were due to determine the effect of parenteral dosage on *M. radiata* [162]. Three levels of adrenaline were given intramuscularly twice a day for 4 weeks. At the end of the experiment there was little to see microscopically and the only vascular lesions were in the aorta. The degree of involvement did not relate to the levels of adrenaline dosage, however, lesions were not seen in the control group. Nor were the lesions preferentially situated at bifurcations or branches. Histological examination revealed sudanophilic lipid within and around intimal foam cells, and there was an increase of stainable mucopolysaccharide. Fibrosis and changes in the elastica were not seen and only occasionally were the lesions large enough to produce elevation of the endothelium. The serum cholesterol, on the other hand, decreased in the groups given adrenaline this fall being largest and more constant in those animals given the highest doses. There was no relationship between the extent of lesions and the degree of reduction of serum cholesterol. It was obviously difficult to assess the effects of endogenous secretion of adrenaline in this work as some of the animals received injections without complaint while others needed considerable restraint. In this regard it is interesting that studies on male and female rhesus monkeys restrained in pillory-type chairs have shown similar falls in serum cholesterol levels [163]. On the whole the fall was greater in the males than the females and took almost 5–10 days to develop. If restraint can be equated to stress then clearly levels of blood catecholamines might be a useful paramter for this. As blood platelets are sensitive to changes in catecholamine levels measurement of their aggregating and release properties in such situations might also be useful. Similar falls in serum cholesterol have been observed in stressed dogs and rats, the fall being particularly obvious in the ester fraction [164], and in man stressful events such as surgery and myocardial infarction have similar effects. This change in sterol level is probably mediated via the hypothalamus and experimental work has shown that hypothalamic changes can affect the levels of circulating lipids and the induction of atherosclerosis [165]. However, these are contradictory reports about changes in serum lipid levels in stress in man and in dogs and rabbits, and about the effects of adrenaline injection in the latter animals, which leaves the question open about the reliability of adrenaline blood levels as an indication of a stressful event.

LORENZEN [166] has provided an extensive review of the biochemical and morphological changes induced by adrenaline and thyroxine. Toxic doses of adrenaline, unlike the effects in *M. radiata,* produce medial calcifications rather like Mönckeberg's sclerosis although occasional intimal changes occurred. In addition he describes an accumulation of mucopolysaccharides in the vessel wall. The effects of thyroxin on the aortic wall were basically the same as those of adrenaline. With both hormones he interprets the changes as initial injury followed by repair. The effects of thyroidectomy was to mollify the changes caused by injections of adrenaline. The net conclusion of this extensive work was that biochemical rather than morphological changes were the most sensitive indicators of vascular injury because the biochemical increases of detectable mucosubstances preceded detectable morphological alterations.

It is often said and many support the notion that atherosclerotic disease is of a multifactorial nature. It is not surprising that though noradrenaline alone may not produce the condition it may, in combination with some other factor such as lipidaemia, cause both atherosclerotic and thrombotic disease. Experiments of this sort have been done by BHATTACHARYA *et al.* [167]. They investigated the effects of a single prolonged infusion of noradrenaline in a dose of 20 mg/kg/mm in normal rhesus monkeys and in animals made lipidaemic and atherosclerotic by feeding 15 g of butter and 1 g of cholesterol per day. Lipidaemia appeared in 4 months and the animals developed moderate atherosclerosis of the aorta and coronary arteries. About 40% of the aortic surface was involved and the lesions were both fatty and fibrous when examined histologically. Three of five animals given noradrenaline developed thrombi in the aorta and coronary arteries. Histological evidence of myocardial infarction was obtained from one animal that survived with ECG changes of ischaemia following noradrenaline infusion. The authors attributed the thrombosis to the effects of the rapid rise of blood pressure that occurred 15 min after the infusion had been started. Thrombi were found only in lipidaemic animals given noradrenaline; none occurred with either factor alone.

In the rat, at least, the most potent hormone affecting the mucopolysaccharide level of the aortic wall is thyroxine. Thyroidectomy and methimazole treatment reduce the level and this effect can be reversed by giving thyroxine. Hypophysectomy does not affect the results so that the effect is directly one of thyroid hormone rather than TSH. Growth hormone specifically raised the chondroitin sulphate concentration in hypophysectomised animals; cortisone specifically lowered the level of chondroitin sulphate. Administration of oestrogen or testosterone or gonadectomy had no effects on the mucopolysaccharide content of the aorta so that thyroxine appeared to be the main hormonal factor in this regard. Whether this situation is also a feature of the non-human primate remains to be seen. It is not possible with our present state of knowledge to provide any useful summary of the relationship between hormones, stress and atherosclerosis in the non-human primate. In the few experiments that have been done on stress [168] the operative factors are many and varied and difficult to assess. For example LAPIN [168] studied the effects of frustration on *Papio hamadryas*. Males and females were housed together then they were separated into adjacent cages and then another male was introduced into the females' cages. The effects upon the frustrated solitary male were hypertensive and led to myocardial infarction. We have already referred to similar observations on the occurrence of myocardial ischaemia in animals at the Philadelphia Zoo. RATCLIFFE attributes these to inappropriate group composition, size and group interaction.

How much caging itself is a form of stress under experimental conditions is difficult to say. We have already seen that serum cholesterol levels rise in captive squirrel monkeys and baboons without dietary provocation, we have seen that adrenaline lowers serum cholesterol. What more can be said?

Inhibition of Experimental Atherosclerosis

Many of the attempts, most of which are fairly recent, that have been made to prevent the induction of atherosclerotic lesions by experiment are based upon the hypotheses of pathogenesis that we have discussed. So we find anti-inflammatory substances being used, hyperoxia being employed and permeability-reducing agents exhibited. We will deal with some of these experiments in this section and others will be included in the next section which is concerned with the regression of lesions after they have formed.

An interesting series of conclusions arises from the work of CHISOLM et al. [169]. They were concerned with the role of hypoxia as an atherogenic factor and in particular the possibility of hypoxia at the blood tissue interface; in this case the arterial intima. They propose that diffusion of oxygen in blood plasma is very sensitive to the level of plasma proteins. They injected albumen and globulin into groups of rabbits and noted intimal changes by electron microscopy. However, they do concede that the changes might be the result of immunological injury rather than due to reduced oxygen diffusion caused by elevated levels of plasma protein. Nevertheless, in a subsequent paper [170] they injected a carotenoid compound crocetin into rabbits fed on an atherogenic diet and showed that the degree of atherosclerosis was reduced and that there was also a 50% fall of plasma cholesterol. The mode of action of crocetin is to increase oxygen diffusion through the plasma. This inhibitory experiment lends support to the popular hypothesis that hypoxia is an important factor in atherogenesis.

Working on the hypothesis that increased permeability of the intima may be a factor in atherogenesis other workers have used pyridinol carbonate (PDC) in an attempt to reduce atherosclerosis in experimental animals. This substance is said to block cellular permeability in the endothelial layer. It is supposed to have an antagonistic effect to bradykinin and in rats it blocks the increase in permeability of vascular endothelium produced by bradykinin, kallikrein, kallidin, RNA and lymph-node-permeability factor. There are reports that the degree of cholesterol-induced atheroma in chickens and rabbits was reduced when PDC was administered. However, MALINOW et al. [171] were unable to show any significant differences in treated and control squirrel monkeys under similar experimental conditions. The same workers used cynomolgus monkeys *(M. irus)* in similar experiments. They confirmed previous studies on the easy induction of coronary disease by dietary means but once again were unable to show any effects of PDC [172]. Results with drugs that affect some part of the inflammatory response have been variable in different species. For example, the antihistamine chlorpheniramine is said to inhibit the development of atherosclerosis in the cholesterol-fed rabbit [173] whereas no such effect could be shown in the pig [174].

Anti-inflammatory agents such as phenylbutazone, amidopyrine [175] and cortisone [176] have inhibitory effects whereas aspirin exacerbates the development of lesions. In general it looks as though the effects of these agents are most often seen in the rabbit where increased endothelial permeability might be assumed to play an important part in atherogenesis, but such conclusions cannot be translated to the non-human primate or to man himself until further work has been done. Techniques of arterial perfusion which we have discussed earlier might find profitable application here.

HOWARD et al. [155], working on the basis of their previous findings that aortic lipase was increased and cholesterol ester hydrolase decreased in the hypercholesterolaemic rabbit, attempted to prevent the development of diet-induced atherosclerosis in baboons by intravenous injections of polyunsaturated phosphatidyl choline (PPC). The idea was to exert the known lipase lowering effect of PPC on the elevated levels that were said to occur in the aortic wall of hypercholesterolaemic baboons. There was some reduction of aortic sudanophilia in the experimental animals but the PPC had no effect on serum cholesterol or phospholipids. Compared to controls the animals given intravenous PPC had normal aortic lipase levels and increased cholesterol ester hydrolase.

Another approach utilised the reported hypolipidaemic effects of orally and parenterally administered chondroitin sulphate (CS), derived from shark fin. If squirrel monkeys which had been fed cholesterol were given CS subcutaneously the hyperlipidaemia and degree of atherosclerosis were less than in experimental subjects not given CS [177]. Reporting again in 1972 MORRISON and BAJWA [178] gave CS to squirrel monkeys every day for 90 days. The material was given intramuscularly and the animals had no other form of treatment. Examination of the extramural branches of the coronary arteries revealed a conspicuous lack of atherosclerosis in the treated group as compared to the controls. This observation is more concerned with regression than with inhibition of lesions. It is certainly difficult to conceive how lesions with a fibrous component can vanish in such a comparatively short space of time.

Regression of Atherosclerotic Lesions

Apart from the previous report by MORRISON and BAJWA [178] no one has reported the total disappearance of atherosclerotic lesions in any animal. It is reasonable to expect that lipid might be induced to issue out of lesions in one way or another but it is likely that scar tissue is immutable to all but surgical procedures.

The possibility of regression is particularly important in relation to the occlusive effects of lipid in atheroma of the coronary circulation and though there is no evidence that experimentally induced atherosclerotic plaques will ever disappear from the large elastic arteries of birds and mammals the ability to reduce their bulk should be valuable. Many human individuals in the 8th and 9th decades of life have severely fibrosed, calcified and even ulcerated coronary arteries but the vessels are dilated with lumina larger than normal. It is the obtruding lump of atheroma that is dangerous. We shall now look at the evidence, mostly from animals other than non-human primates for the regression of such lesions.

It is well known and we have already discussed in a previous section that flux of free cholesterol from cells to serum increases as the rate of esterification of free cholesterol increases. This phenomenon has been demonstrated to occur in human arteries and is mediated by the enzyme LCAT [179]. Here then is a possible mechanism for the egress of cholesterol from the vessel wall and possibly also from atherosclerotic lesions.

CONSTANTINIDES [142] as a result of intensive comprehensive studies on cholesterol atherogenesis in the rabbit conceded the possibility that lipid may be removable from early fatty streak-type lesions but emphasises that once fibrosis has set in the lesion, including its lipid content, is irreversible. He showed, for example, that when lesions were produced in rabbits after 2 months feeding, the diets were then returned to normal and the animals were killed 2 years later, the lesions were bigger and more fibrous than in animals killed early in the experiment [180]. Experiments such as these hold little hope for regression of primate atherosclerosis and the work of MARUFFO and PORTMAN [39] in the squirrel monkey tends to confirm this view. They fed an atherogenic diet to squirrel monkeys for 3 months then followed this with a control diet for a further 3 or 5 months. These animals showed coronary lesions which certainly did not progress as occurred with monkeys fed the diet for a longer time, nor did they regress to the condition found in animals that have received control diet alone. In these experiments, however, it is likely that regression may have taken place if more time had allowed on basal diet.

Encouraged by the several reported decreases of atherosclerosis that were purported to occur in human wasting diseases ARMSTRONG and MEGAN [182] pursued their extensive studies of regression of lesions in rhesus monkeys. Up to 1970 no one had established whether the withdrawal of an atherogenic stimulus to a non-human primate had any effect on the atheromatous lesions and whether the lumen enlarged to normal size after regression. There were several groups of animals in their experiment but basically the scheme was to feed a control diet for 6 weeks, an atherogenic diet for 17 months, then the regression diets for 40 months. These regression diets were either linoleate-rich or low-fat in type. The atherogenic diet as might be expected produced prominent stenosing coronary atheroma. Those given the regression diets had smaller fibrotic lesions with scantily stainable lipid. The effects of the two sorts of regression diets were much the same. The increase in luminal size of the arteries of animals on regression diet was shown not to be due to dilatation of the vessel because the intimal area of the atherosclerotic monkey arteries was more than three times greater than that of the arteries from animals in the regression experiment.

In a similar series of experiments [182] the same workers set out to determine the chemical changes that occurred in arteries during regression. Their original view based on morphology, namely that lipid comes out of the lesion in regression, was confirmed. The free cholesterol level of the lesions fell by 53% and the ester cholesterol by 69% as compared to controls. In order to exclude a possible dilution effect by reparative tissue in the regression studies they established that the tissue concentration of cholesterol and its esters was much lower in the lesions in regression. So far as other lipids were concerned no changes in triglycerides or phospholipids were observed on regression though the levels of these lipids were increased in lesions.

The main component that emerged from the artery appeared to be free cholesterol and some ester stayed behind as a 'biochemical scar of the atherosclerotic process'. Most of the lipid efflux had, as far as they could tell, occurred in the first half of the regression period that is to say in about 20 months. The lipid that remained was deeper in the lesions; that which emerged from the vessel was more superficial in the intima. It does seem from this work that the age of the lesions and the situation of the lipid may be important factors in determining whether regression occurs or not and it may be that the rabbit lesions of CONSTANTINIDES [142] might have been too well established for lipid to get out.

Human atheromas do show evidence of equilibration of their contained cholesterol with plasma cholesterol in studies using labelled cholesterol so at least a chance of lipid removal does exist in human disease [183].

TUCKER et al. [184] devised experiments in rhesus monkeys to investigate the reversibility of early lesions. The diet was fed for 8 weeks only. Their result showed that aortic lesions can be produced predictably in 8 weeks and that changes in these lesions occurred 16 weeks after withdrawal from the diet. The amount of lipid was much less in the lesions of the regression group. In the non-regression group the intimal lesions consisted of smooth muscle cells and monocytes both contained lipid. The monocytes; however, appeared to be degenerating and spilling lipid into the extracellular space. In the regression lesions monocytes were much less numerous and the smooth muscle cells, though present, contained much less lipid. If anything there was relatively more extracellular lipid in regression lesions which presumably meant that the monocytes had broken down in the regression phase liberating cholesterol which then equilibrated with cholesterol in the plasma. This being so lipid could emerge from the vessel wall by the processes that we have already discussed.

There is clearly a good deal of confusion about the question of reversibility of rabbit atheroma. FRIEDMAN and BYERS [185] like CONSTANTINIDES [142] maintain that the lesions are irreversible. As we have already said this may be related to the age of the lesion. HORLICK and KATZ [186] in one of the earliest regression studies showed regression of coronary lesions in the chick and PETERSON and HIRST [187] made the important observation that the early lesions could be induced to regress but not the older fibrotic ones. In a 10-week study KJELDSEN et al. [188] exposed rabbits already fed a 2% cholesterol diet for 5 weeks to hyperoxia (28% oxygen). The degree of visible aortic atheromatosis and the aortic content of cholesterol, phospholipid and triglyceride was significantly lower in the group subjected to hyperoxia. Another factor to be considered in studies on regression is the effect of the application of multiple methods to induce regression. In addition to hyperoxia alone others have tried low fat diets coupled with hyperoxia, cholestyramine and oestrogen [189]. Atheroma was induced in 14 weeks by feeding a high cholesterol diet, then, 10 weeks regression phase. One had low fat alone, and one had low fat and hyperoxia with either cholestyramine or oestrogen or both of these substances. In general, the combination of drugs was the most effective in inducing regression: low-fat diets alone had little effect.

When lesions were induced by injections of bovine serum albumen (BSA) coupled with cholesterol feeding, reversibility was observed only in the group given BSA alone [153]. When the two treatments were given irreversible lesions occurred in as short a time as 2 weeks. To some extent then, the reversibility or otherwise of experimentally induced atherosclerosis in the rabbit depends not only upon the age of the lesions but also upon the particular regime that was used to produce them.

Experiments designed to observe the sequential changes in individual plaques when regression diets are imposed are very few. DE PALMA et al. [190] watched the changes in plaques in the inferior mesenteric artery of dogs made severely hypercholesterolaemic by thyroid ablation and cholesterol feeding. If the elevated cholesterol levels are maintained for at least 12 months, ulceration, calcification and haemorrhage are found in the atherosclerotic plaques. In general they found that lesions could be induced to regress and that this depended mainly upon the duration and level of the hypercholesterolaemia. They were able to watch related lesions in the inferior mesenteric arteries in animals on a regression diet by means of repeated laparotomy.

In conclusion then regression of lesions is possible; it seems to involve the efflux of superficial free cholesterol from the plaques. The older the lesions and the deeper the lipid in the vessel wall, the less likely is regression to occur. No one has shown disappearance of the fibrous components of coronary plaques in any study.

Conclusion

It is a most remarkable fact that within the primate order only one genus, man, is prone to develop atherosclerotic disease that embarrasses the circulation to various organs and that only in man is thrombosis a complication of such disease. All other members of the order show lesions of various degrees in various vessels but in all they are trivial and uncompromising. With suitable forms of provocation the disease in non-human primates can be augmented and in some cases resemblance to the human situation is a close one. The reasons for the unique behaviour of man in this regard must be explained by various studies that have and should be made in non-human primates. They are likely to provide the most fruitful sources of information in the experimental field of atherogenesis.

References

1   GOLDBLATT, H.: Studies on experimental hypertension. III. The production of persistent hypertension in monkeys (macaque) by renal ischemia. J. exp. Med. *65:* 671–675 (1937).
2   KAWAMURA, R.: in Neue Beiträge zur Morphologie und Physiologie der Cholinesterinsteatose, p. 267 (Fischer, Jena 1927).

3   Taylor, C.B.; Nelson-Cox, L.G.; Hall-Taylor, B.J., and Cox, G.E.: Atherosclerosis in monkeys with moderate hypercholesterolemia induced by dietary cholesterol. Fed. Proc. Fed. Am. Socs exp. Biol. *16:* 374 (1957).
4   Sperry, W.M.; Jailer, J.W., and Engle, E.T.: The influence of diet on the cholesterol concentration of the blood serum in normal, spayed and hypothyroid monkeys. Endocrinology *35:* 38–48 (1944).
5   Lofland, H.B., jr.; Clarkson, T.B.; St. Clair, R.W., and Lehner, N.D.M.: Studies on the regulation of plasma cholesterol levels in squirrel monkeys of two genotypes. J. Lipid. Res. *13:* 39–47 (1972).
6   Strong, J.P. and McGill, H.C., jr.: Diet and experimental atherosclerosis in baboons. Am. J. Path. *50:* 669–690 (1967).
7   Cox, G.E.; Nelson, L.G., and Taylor, C.B.: Hypercholesterolemia in monkeys ingesting an adequate diet. Circulation *10:* 38 (1954).
8   Taylor, C.B.; Cox, G.E.; Counts, M., and Yogi, N.: Fatal myocardial infarction in the rhesus monkey with diet-induced hypercholesterolemia. Am. J. Path. *35:* 674 (1959).
9   Taylor, C.B.; Cox, G.E.; Manalo-Estrella, P., and Southworth, J.: Atherosclerosis in rhesus monkeys. Archs Path. *74:* 16–34 (1962).
10 Mann, G.V. and Andrus, S.B.: Xanthomatosis and atherosclerosis produced by diet in an adult rhesus monkeys. J. Lab. clin. Med. *48:* 533–550 (1956).
11 Armstrong, M.L.; Connor, W.E., and Warner, E.D.: Xanthomatosis in rhesus monkeys fed a hypercholesterolemic diet. Archs Path. *84:* 227–237 (1967).
12 Armstrong, M.L. and Warner, E.D.: Morphology and distribution of diet-induced atherosclerosis in rhesus monkeys. Archs Path. *92:* 395–401 (1971).
13 Schwartz, C.J. and Mitchell, J.R.A.: Cellular infiltration of the human arterial adventitia associated with atheromatous plaques. Circulation *26:* 73–78 (1962).
14 Manning, P.J. and Clarkson, T.B.: Development, distribution and lipid content of diet-induced atherosclerotic lesions of rhesus monkeys. Expl molec. Path. *17:* 38–54 (1972).
15 Scott, R.F.; Jones, R.; Daoud, A.S.; Zumbo, O.; Coulstom, F., and Thomas, W.A.: Experimental atherosclerosis in rhesus monkeys. Expl molec. Path. *7:* 34–57 (1967).
16 Gresham, G.A. and Howard, A.N.: The independent production of thrombosis and atherosclerosis in the rat. Br. J. exp. Path. *42:* 166–170 (1961).
17 Vesselinovitch, D.; Getz, G.S.; Hughes, R.H., and Wissler, R.W.: Atherosclerosis in the rhesus monkey fed three food fats. Atherosclerosis *20:* 303–321 (1974).
18 Geer, J.C.: Fine structure of human aortic intimal thickening and fatty streaks. Lab. Invest. *14:* 1764–1783 (1965).
19 Haust, M.D. and More, R.H.: The role of differentiating smooth muscle cells in the organisation of the human aorta. An electron microscopic study. Fed. Proc. Fed. Am. Socs exp. Biol. *25:* 475 (1966).
20 McGill, H.C.: Fatty streaks in the coronary arteries and aorta. Lab. Invest. *18:* 560–564 (1968).
21 Kramsch, D.M. and Hollander, W.: Occlusive atherosclerotic disease of the coronary arteries in monkey *(Macaca irus)* produced by diet. Expl molec. Path. *9:* 1–22 (1968).

22 BÄURLE, W.: Die Coronarsklerose bei Hypertonie. Beitr. path. Anat. *111:* 108–124 (1950).
23 MOON, H.D. and RINEHART, J.F.: Histogenesis of coronary atherosclerosis. Circulation *6:* 481–488 (1952).
24 VLADOVER, Z.; ABRAMOVICI, A.; NEUFELD, H.N., and LIBAN, E.: Coronary arteries in Yemenites. J. Atheroscler. Res. *7:* 161–170 (1967).
25 TOOR, M.; KATCHALSKY, A., AGMON, J., and ALLALOUF, D.: Atherosclerosis and related factors in immigrants to Israel. Circulation *22:* 265–279 (1960).
26 LEMAIRE, A. et COTTET, J.: Le syndrome biochimique de l'athérosclérose animale. Presse méd. *63:* 1339–1341 (1955).
27 YOUNGER, R.K.; SCOTT, W.; BUTTS, W.H., and STEPHENSON, S.E.: Rapid production of experimental hypercholesterolemia and atherosclerosis in the rhesus monkey. J. surg. Res. *9:* 263–271 (1969).
28 HOWARD, A.N.; GRESHAM, G.A.; RICHARDS, C., and BOWYER, D.E.: Serum proteins, lipoproteins and lipids in baboons given normal and atherogenic diets; in VAGTBORG The baboon in medical research, vol. 1 (Texas University Press, Austin 1965).
29 KRITCHEVSKY, D.; SHAPIRO, I.L., and WERTHESSEN, M.T.: Biosynthesis of cholesterol in the baboon. Biochim. biophys. Acta *65:* 537–586 (1962).
30 HOWARD, A.N.; GRESHAM, G.A.; HALES, C.N.; LINDGREN, F.T., and KATZBERG, A.H.: Atherosclerosis in baboons: pathological and biochemical studies; in The baboon in medical research, vol. 2, pp. 333–350 (Texas University Press, Austin 1965).
31 FOY, H.; KONDI, A., and MBAYA, V.: Haematologic and 'biochemical' indices in the East African baboon. Blood *26:* 682–686 (1965).
32 GRESHAM, G.A.; HOWARD, A.N.; MCQUEEN, J., and BOWYER, D.E.: Atherosclerosis in primates. Br. J. exp. Path. *46:* 94–103 (1965).
33 STRONG, J.P.; EGGEN, D.A.; NEWMAN, W.P., III, and MARTINEZ, R.D.: Naturally occurring and experimental atherosclerosis in primates. Ann. N.Y. Acad. Sci. *149:* 882–894 (1968).
34 CHOBANIAN, A.V.; BURROWS, B.A., and HOLLANDER, W.: Body cholesterol metabolism in man. II. Measurement of the body cholesterol miscible pool and turnover rate. J. clin. Invest. *41:* 1738–1744 (1962).
35 MALINOW, M.R.; MARUFFO, C.A., and PERLEY, A.M.: Atherosclerosis in monkeys *(Saimiri sciurea).* J. Path. Bact. *92:* 491–510 (1966).
36 GORE, I.; ROBERTSON, W.B.; HIRST, A.E.; HADLEY, G.G., and KOSEKI, Y.: Geographic differences in the severity of aortic and coronary atherosclerosis. Am. J. Path. *36:* 559–574 (1960).
37 KRITCHEVSKY, D. and MCCANDLESS, R.F.J.: Weekly variations in serum cholesterol levels of monkeys. Proc. Soc. exp. Biol. Med. *95:* 152–154 (1957).
38 MIDDLETON, C.C.; CLARKSON, T.B.; LOFLAND, H.B., and PRICHARD, R.W.: Diet and atherosclerosis in squirrel monkeys. Archs Path. *83:* 145–153 (1967).
39 MARUFFO, C.A. and PORTMAN, O.W.: Nutritional control of coronary artery atherosclerosis in the squirrel monkey. J. Atheroscler. Res. *8:* 237–247 (1968).
40 PORTMAN, O.W. and ANDRUS, S.B.: Comparative evaluation of three species of New World monkeys for studies of dietary factors, tissue lipids and atherogenesis. J. Nutr. *87:* 429–438 (1965).

41 MANN, G.V.; ANDRUS, S.B.; MCNALLY, A., and STARE, F.J.: Experimental atherosclerosis in cebus monkeys. J. exp. Med. 98: 195–218 (1953).
42 WISSLER, R.W.; FRAZIER, L.E.; HUGHES, R.H., and RASMUSSEN, R.A.: Atherogenesis in the cebus monkey. I. A comparison of three food fats under controlled dietary conditions. Archs Path. 74: 312–322 (1962).
43 BULLOCK, B.C.: Effect of age and diet on coronary artery atherosclerosis in Cebus albifrons. Fed. Proc. Fed. Am. Socs exp. Biol. 26: 371 (1967).
44 MACNINTCH, J.E.; ST. CLAIR, R.W.; LEHNER, N.D.M.; CLARKSON, T.B., and LOFLAND, H.B.: Cholesterol metabolism and atherosclerosis in cebus monkeys in relation to age. Lab. Invest. 16: 444–452 (1967).
45 NEWMAN, W.P., III; EGGEN, D.A., and STRONG, J.P.: Comparison of arterial lesions and serum lipids in spider and rhesus monkeys on an egg and butter diet. Atherosclerosis 19: 75–86 (1974).
46 WISSLER, R.W.; HUGHES, R.H.; FRAZIER, L.E.; GETZ, G.S., and TURNER, D.F.: Aortic lesions and blood lipids in rhesus monkeys fed 'table-prepared' human diets. Circulation 32: suppl. 2, pp. 220–221 (1965).
47 KOTZE, J.P.; NEUHOFF, J.S.; ENGELBRECHT, G.P.; MERWE, G.J. VAN DER; DU PLESSIS, J.P., and HORN, L.P.: The fatty acid composition of cholesteryl esters, phospholipids and triglycerides of the baboon, Papio ursinus. Atherosclerosis 19: 469–476 (1974).
48 COREY, J.E. and HAYES, K.C.: Effect of diet on hepatic and intestinal lipogenesis in squirrel, cebus and cynomolgus monkeys. Atherosclerosis 20: 405–416 (1974).
49 KRITCHEVSKY, D.; DAVIDSON, L.M.; SHAPIRO, I.L.; KIM, H.K.; KITAGAWA, M.; MALLIOTRA, S.; NAIR, P.P.; CLARKSON, T.B.; BERSOHN. I., and WINTER, P.A.D.: Lipid metabolism and experimental atherosclerosis in baboons: influence of cholesterol-free, semi-synthetic diets. Am. J. clin. Nutr. 27: 29–50 (1974).
50 ARMSTRONG, M.L.; MEGAN, M.B., and WARNER, E.D.: Intimal thickening in normocholesterolemic rhesus monkeys fed low supplements of dietary cholesterol. Circulation Res. 34: 447–454 (1974).
51 DREIZEN, S.; LEVY, B.M., and BERNICK, S.: Diet-induced atherosclerosis in the Marmo set. Proc. Soc. exp. Biol. Med. 143: 1218–1223 (1973).
52 EGGEN, D.A.; STRONG, J.P., and NEWMAN, W.P., III: Experimental atherosclerosis in primates: a comparison of selected species. Ann. N.Y. Acad. Sci. 162: 80–88 (1969).
53 MANNING, P.J.; CLARKSON, T.B., and LOFLAND, H.B.: Cholesterol absorbtion, turnover and excretion rates in hypercholesterolemic rhesus monkeys. Expl molec. Path. 14: 75–89 (1971).
54 KAPLAN, J.A.; COX, G.E., and TAYLOR, C.B.: Cholesterol metabolism in man: studies on absorbtion. Archs Path. 76: 359–368 (1963).
55 FRASER, R.; DUBIEN, L.; FOSSLIEN, E., and WISSLER, R.W.: Transport of cholesterol in thoracic duct lymph and serum of rhesus monkeys fed cholesterol with various food fats. Atherosclerosis 16: 203–216 (1972).
56 RAO, B.S.N.; RAO, A.R., and GOPALAN, C.: Study of ^{131}I triolein tolerance curves in different species of animals. J. Atheroscler. Res. 6: 447–454 (1966).
57 KAO, V.C.Y. and WISSLER, R.W.: A study of the immunohistochemical localisation of serum lipoproteins and other plasma proteins in human atherosclerotic lesions. Expl molec. Path. 4: 465–479 (1965).

58 AMENTA, J.S. and WATERS, L.L.: The precipitation of serum lipoproteins by mucopolysaccharides extracted from aortic tissue. Yale J. Biol. Med. *41:* 1732 (1962).
59 SMITH, E.B.; SLATER, R.S., and CHU, P.K.: The lipids in raised fatty and fibrous lesions in human aorta. J. Atheroscler. Res. *8:* 399–420 (1968).
60 MACMILLAN, R.; ADAMS, C.W.M., and IBRAHIM, M.Z.M.: Histochemical identification of plasma proteins in the human aortic intima. J. Path. Bact. *89:* 226–231 (1965).
61 GÉRO, S.J.; GERGELY, L.; JAKAB, L.; SZÉKELY, J., and VIRÁG, S.: Comparative immunoelectrophoretic studies on homogenates of aorta, pulmonary arteries and inferior vena cava of atherosclerotic individuals. J. Atheroscler. Res. *1:* 88–91 (1961).
62 WALTON, K.W. and WILLIAMSON, N.: Histological and immunofluorescent studies on the evolution of the human atherosclerotic plaque. J. Atheroscler. Res. *8:* 599–624 (1968).
63 SCOTT, P.J. and HURLEY, P.J.: The distribution of radio-iodinated serum albumen and low-density lipoproteins in tissues and the arterial wall. Atherosclerosis *11:* 77–103 (1970).
64 BLATON, V.; VANDAMME, D.; DECLERCQ, B.; VASTESAEGER, M.; MORTELMANS, J., and PEETERS, H.: Dietary induced hyperbetalipoproteinemia in chimpanzees: comparison to the human hyperlipoproteinemia. Expl molec. Path. *20:* 132–146 (1974).
65 STARE, F.J.; ANDRUS, S.B., and PORTMAN, O.W.: Primates in medical research with special reference to New World monkeys; in PICKERING Proc. Conf. on Research with Primates, No. 59 (Tektronix Foundation, Beaverton).
66 MANN, G.V.; ANDRUS, S.B.; MCNALLY, A., and STARE, F.J.: Diet and cholesterorolemia in chimpanzees. Fed. Proc. Fed. Am. Socs exp. Biol. *22:* 642 (1963).
67 DAYTON, S. and HASHIMOTO, S.: Cholesterol flux and metabolism in arterial tissue and in atheromata. Expl molec. Path. *13:* 253–268 (1970).
68 CONSTANTINIDES, P.: Lipid deposition in injured arteries. Electron microscopic study. Archs Path. *85:* 106–112 (1968).
69 PORTMAN, O.W. and ALEXANDER, M.: Lipid compostion of aortic intima plus media and other tissue functions from foetal and adult rhesus monkeys. Archs Biochem. Biophys. *117:* 357–365 (1966).
70 SMITH, E.B.; EVANS, P.H., and DOWNHAM, M.D.: Lipid in the aortic intima the correlation of morphological and chemical characteristics. J. Atheroscler. Res. *7:* 171–186 (1967).
71 PORTMAN, O.W.: Atherosclerosis in non-human primates: sequences and possible mechanisms of change in phospholipid composition and metabolism. Ann. N.Y. Acad. Sci. *162:* 120–136 (1969).
72 DERVICHIAN, D.G.: in DANIELLI, PANKHURST and RIDDIFORD Surface phenomenon in chemistry and biology, p. 70 (Pergamon Press, New York 1958).
73 DAYTON, S.; HASHIMOTO, S., and JESSAMY, J.: Cholesterol kinetics in the normal rat aorta, and the influence of different types of dietary fat. J. Atheroscler. Res. *1:* 444–460 (1961).
74 LOFLAND, H.B.; MOURY, D.M.; HOFFMAN, C.W., and CLARKSON, T.B.: Lipid metabolism in pigeon aorta during atherogenesis. J. Lipid Res. *6:* 112–118 (1965).
75 ABDULLA, Y.H.; ORTON, C.C., and ADAMS, C.W.M.: Cholesterol esterification by transacylation in human and experimental atheromatous lesions. J. Atheroscler. Res. *8:* 967–973.

76 RAO, A.R. and RAO, B.S.M.: Incorporation of C 1–^{14}C) acetate into the lipids of aortas of different species. J. Atheroscler. Res. *8:* 59–67 (1968).
77 BOWYER, D.E.; HOWARD, A.N.; GRESHAM, G.A.; BATES, D., and PALMER, B.V.: Aortic perfusion in experimental animals: a system for study of lipid synthesis and accumulation. Prog. biochem. Pharmacol., vol. *4:* 235–243 (Karger, Basel 1968).
78 PATELSKI, J.; BOWYER, D.E.; HOWARD, A.N., and GRESHAM, G.A.: Changes in phospholipase A, lipase and cholesterol ester activity in the aorta in experimental atherosclerosis in rabbit and rat. J. Atheroscler. Res. *8:* 221–228 (1968).
79 LOFLAND, H.B. and CLARKSON, T.B.: Certain metabolic patterns of atheromatous pigeon aortas. Archs Path. *80:* 291–296 (1965).
80 LOFLAND, H.B.; ST. CLAIR, R.W.; CLARKSON, W.B.; BULLOCK, B.C., and LEHMER, N.D.M.: Atherosclerosis in cebus monkeys. II. Arterial metabolism. Expl molec. Path. *9:* 57–70 (1968).
81 ST. CLAIR, R.W.; LOFLAND, H.B., and CLARKSON, T.B.: Influence of atherosclerosis on the composition, synthesis, and esterification of lipids in aortas of squirrel monkeys *(Saimiri sciurea)*. J. Atheroscler. Res. *10:* 193–206 (1969).
82 STOKKE, K.T. and NORUM, K.R.: Deterioration of lecithin cholesterol acyltransferase in human blood plasma. Scand. J. clin. Lab. Invest. *27:* 21–27 (1971).
83 LACKO, A.G.; RUTENBERG, H.L., and SOLOFF, L.A.: Serum cholesterol esterification in species resistant and susceptible to atherosclerosis. Atherosclerosis *19:* 297–305 (1974).
84 SWELL, L.; FIELD, H., and TREADWELL, C.R.: Correlation of arachidonic acid of serum cholesterol esters in different species with suceptibility to atherosclerosis. Proc. Soc. exp. Biol. Med. *104:* 325–328 (1960).
85 BONNER, M.J.; MILLER, B.F., and KOTHARI, H.V.: Lysosomal enzymes in aortas of species susceptible and resistant to atherosclerosis. Proc. Soc. exp. Biol. Med. *139:* 1359–1362 (1972).
86 JONES, D.C.; LOFLAND, H.B.; CLARKSON, T.B., and ST. CLAIR, R.W.: Interaction of diet and phenotypes on plasma cholesterol concentration in squirrel monkeys. Circulation *45/46:* suppl. 2, p. 19 (1972).
87 STOKKE, K.T.: Cholesteryl ester metabolism in liver and blood plasma of various animal species. Atherosclerosis *19:* 393–406 (1974).
88 HASHIMOTO, S. and DAYTON, S.: Cholesterol-esterifying activity of aortas from atherosclerosis-resistant and atherosclerosis-susceptible species. Proc. Soc. exp. Biol. Med. *145:* 89–92 (1974).
89 GERTLER, N.M. and OPPENHEIMER, B.S.: Total cholesterol lipid phosphorous ratio. Its significance in atherosclerosis. Geriatrics *9:* 157 (1954).
90 PARKER, F.; ORMSBY, J.W.; PETERSON, N.F.; ODLAND, G.F., and WILLIAMS, R.H.: *In vitro* studies of phospholipid synthesis in experimental atherosclerosis. Possible role of myointimal cells. Circulation Res. *19:* 700–710 (1966).
91 NEWMAN, H.A.I. and ZILVERSMIT, D.B.: Accumulation of lipid and nonlipid constituents in rabbit atheroma. J. Atheroscler. Res. *4:* 261–271 (1964).
92 BORENSZTAJN, J.; GETZ, G.S., and WISSLER, R.W.: The *in vitro* incorporation of (^{3}H) thymidine into DNA and of ^{32}P into phospholipids and RNA in the aorta of rhesus monkeys during early atherogenesis. Atherosclerosis *17:* 269–280 (1973).

93 PEETERS, H.; BLATON, V.; DECLERCQ, B.; HOWARD, A.N., and GRESHAM, G.A.: Lipid changes in the plasma lipoproteins of baboons given an atherogenic diet. Atherosclerosis *12:* 283–290 (1970).
94 VERESS, B.; BÁLINT, A.; KÓCZÉ, A.; NAGY, Z., and JELLINEK, H.: Increasing aortic permeability by atherogenic diet. Atherosclerosis *11:* 369–371 (1970).
95 SOMER, J.B.; BELL, F.P., and SCHWARTZ, C.J.: Focal differences in lipid metabolism in the young pig aorta. Atherosclerosis *20:* 11–21 (1974).
96 MALYUK, W.J.: Autoradiographic investigation of ^{35}S incorporation into the arterial wall of rats in post-natal atherogenesis. Proc. 8th Scientific Conf. on Gerontology, etc., vol. 2, p. 117–118 (Pedagogiko, Moscow 1971).
97 KASABIJAM, S.S.: The histochemistry of hyaluronic acid in the aorta in atherosclerosis. Kardiologiya *4:* 18 (1964).
98 ROSSI, G.B.; CACCAVOLE, E., and PARAZZI, L.C.: Salicylate and increased vascular permeability due to hyaluroniduse. Nature, Lond. *200:* 685–686 (1963).
99 KLYNSTRA, F.B.: On the passage-restricting role of acid mucopolysaccharides in the endothelium of pig aortas. Atherosclerosis *19:* 215–220 (1974).
100 LEHNINGER, A.L.: Cell counts and ground substance in animal tissues; in LEHNINGER Biochemistry, p. 235 (Worth, New York 1970).
101 SIMKIN, P.A.: Permeability of aortic endothelium. Lancet *i:* 793 (1972).
102 GOLDBLATT, H.: Studies on experimental hypertension. III. The production of persistent hypertension in monkeys (Macacque) by renal ischemia. J. exp. Med. *65:* 671–675 (1937).
103 MCGILL, H.C., jr.; FRANK, M.H., and GEER, J.C.: Aortic lesions in hypertensive monkeys. Archs Path. *71:* 96–102 (1961).
104 MASSOM, G.M.C.; PLAHL, G.; CORCORAN, A.C., and PAGE, I.H.: Accelerated hypertensive vascular disease from saline and renin in nephrectomised dogs. Archs Path. *55:* 85–97 (1953).
105 PRATHRAP, K.: Morphology of two year old healed platelet rich thrombi in the femoral arteries of normocholesterolemic monkeys. J. Path. Bact. *110:* 145–151 (1973).
106 PRATHAP, K.: Natural history of platelet rich mural thrombi in systemic arteries of hypercholesterolemic monkeys. Light and electron microscopic observation. J. Path. Bact. *110:* 203–212 (1973).
107 COHEN, P. and MCCOMBS, H.L.: Platelets and atherosclerosis. I. Augmentation of cholesterol atherogenesis in the rabbit by a phlebotomy programme designed to induce thrombocytosis. Br. J. exp. Path. *48:* 346–356 (1967).
108 GORDON, J.L.: Personal commun. (1974).
109 MYASNIKOV, A.L.: Influence of some factors on development of experimental cholesterol atherosclerosis. Circulation *17:* 99–113 (1958).
110 FILLIOS, L.C.; ANDRUS, S.B., and NAITO, C.: Coronary lipid deposition during chronic anemia or high altitude exposure. J. appl. Physiol. *16:* 103–106 (1961).
111 KJELDSEN, K.; WANSTRUP, J., and ASTRUP, P.: Enhancing influence of arterial hypoxia on the development of atheromatosis in cholesterol fed rabbits. J. Atheroscler. Res. *8:* 835–845 (1968).
112 LORENZEN, I. and HELIN, P.: Arteriosclerosis induced by hypoxia. Acta path. microbiol. scand. *69:* 158–159 (1967).

113 ASTRUP, P.; KJELDSEN, K., and WANSTRUP, J.: Enhancing influence of carbon monoxide on the development of atheromatosis in cholesterol-fed rabbits. J. Atheroscler. Res. *7:* 343–354 (1967).
114 WEBSTER, W. S.; CLARKSON, T. B., and LOFLAND, H. B.: Carbon monoxide-aggravated atherosclerosis in the squirrel monkey. Expl molec. Path. *13:* 36–50 (1958).
115 HELLUNG-LARSEN, P.; LAUSEN, T.; KJELDSEN, K., and ASTRUP, P.: Lactate dehydrogenase isoenzymes of aortic tissue in rabbits exposed to carbon monoxide. J. Atheroscler. Res. *8:* 343–349 (1968).
116 LOOMEIJER, F.J. and OSTENDORF, J.P.: Oxygen consumption of the thoracic aorta of normal and hypercholesterolemic rats. Circulation Res. *7:* 466–467 (1959).
117 LAZZARINI-ROBERTSON, A., jr.: Effects of oxygen tension on the uptake of labeled lipoproteins by human atheromatous plaques. Circulation *24:* 1096–1097 (1961).
118 BULLOCK, B.C.; CLARKSON, T.B.; LEHNER, N.D.M.; LOFLAND, H.B., and ST. CLAIR, R.W.: Atherosclerosis in *Cebus albifrons* monkeys. III. Clinical and pathological studies. Expl molec. Path. *10:* 39–62 (1969).
119 THOMSEN, H.K.: Carbon monoxide-induced atherosclerosis in primates. An electron-microscopic study on the coronary arteries of *Macaca irus* monkeys. Atherosclerosis *20:* 233–240 (1974).
120 GRESHAM, G.A. and HOWARD, A.N.: in BOURNE Non-human primates in medical research, p. 237 (Academic Press, New York 1973).
121 SALANS, L.G.; KNITTLE, J.L., and HIRSCH, J.J.: The role of adipose cell size and adipose tissue insulin in the carbohydrate intolerance of human obesity. J. clin. Invest. *47:* 153–165 (1968).
122 PETERS, M. and HALES, C.N.: Plasma insulin concentration after myocardial infarction. Lancet *i:* 1144–1145 (1965).
123 HAMILTON, C.L.; KUO, P.T., and FENG, L.Y.: Experimental production of syndromes of obesity hyperinsulinemia and hyperlipidemia in monkeys. Proc. Soc. exp. Biol. Med. *140:* 1005–1008 (1972).
124 HOWARD, A.N.; GRESHAM, G.A.; HALES, C.N.; LINDGREEN, F.T., and KATZBERG, A.A.: Atherosclerosis in baboons: pathological and biochemical studies; in The baboon in medical research, vol. 2 (Southwest Foundation for Research and Education, San Antonio 1963).
125 LEHNER, N.D.M.; CLARKSON, T.B., and LOFLAND, H.B.: The effect of insulin deficiency, hypothyroidism and hypertension on atherosclerosis in the squirrel monkey. Expl molec. Path. *15:* 230–244 (1971).
126 GILLMAN, J.; GILBERT, C., and ALLAN, J.C.: The relationship of hyperglycemia to hyperlipemia and ketonemia in depancreatised baboons *(Papio ursinus)*. J. Endocr. *17:* 349–362 (1958).
127 FISHER, R.E.: in BREST and MOYER Atherosclerotic vascular disease, pp. 194–206 (Appleton Century Crofts, New York 1967).
128 EPSTEIN, F.H.; OSTRANDER, L.D., jr.; JOHNSON, B.C.; PAYNE, M.W.; HAYNER, N.S.; KELLER, J.B., and FRANCIS, T., jr.: Epidemiological studies of cardiovascular disease in a total community – Tecumseh, Michigan. Ann. intern. Med. *62:* 1170–1187 (1965).
129 COREY, J.E.; HAYES, K.C.; DORR, B., and HEGSTED, D.M.: Comparative lipid response of four primate species to dietary changes in fat and carbohydrate. Atherosclerosis *19:* 119–134 (1974).

130 PUCAK, G.H.; LEHNER, N.D.M.; CLARKSON, T.B.; BULLOCK, B.C., and LOFLAND, H.B.: Spider monkeys (*Ateles* sp.) as animal models for atherosclerosis research. Expl molec. Path. *18:* 32–49 (1973).
131 MACNINTCH, J.E.; ST. CLAIR, R.W.; LEHNER, N.D.M.; CLARKSON, T.B., and LOFLAND, H.B.: Cholesterol metabolism and atherosclerosis in cebus monkeys in relation to age. Lab. Invest. *16:* 444–452 (1967).
132 MIDDLETON, C.C.; CLARKSON, T.B.; LOFLAND, H.B., and PRICHARD, R.W.: Diet and atherosclerosis in squirrel monkeys. Archs Path. *83:* 145–153 (1967).
133 COLTART, T.M.: Changes in serum phospholipids in male and female baboons on a sucrose diet. Nature, Lond. *222:* 575–576 (1969).
134 RINEHART, J.F. and GREENBERG, L.D.: Vitamin B_6 deficiency in the rhesus monkey. Am. J. clin. Nutr. *5:* 318–328 (1956).
135 MANN, G.V.; WATSON, P.L.; MCNALLY, A.; GODDARD, J., and STARE, F.J.: Experimental atherosclerosis in cebus monkeys. J. exp. Med. *98:* 196–218 (1953).
136 GREENBERG, L.D.; MISCONI, L.Y., and WANG, A.C.: Desmosine and isodesmosine levels of aortic elastin in control and pyridoxine-deficient monkeys. Res. Commun. chem. Path. Pharmac. *2:* 869–875 (1971).
137 MANN, C.V.; ANDRUS, S.B.; MCNALLY, A., and STARE, F.J.: Experimental atherosclerosis in cebus monkeys. J. exp. Med. *98:* 196–218 (1953).
138 HARTROFT, W.S. and THOMAS, W.A.: in SANDLER and BOURNE Atherosclerosis and its origin, p. 445 (Academic Press, New York 1963).
139 HUNT, R.D.; GARCIA, F.G., and HEGSTED, D.M.: Hypervitaminosis D in New World monkeys. Am. J. clin. Nutr. *22:* 358–366 (1969).
140 NORMAN, A.W.: The mode of action of vitamin D. Biol. Rev. *43:* 97–137 (1968).
141 KIRK, J.E.: Vitamin contents of arterial tissue. Monogr. Atheroscler., vol. 3, pp. 4–11 (Karger, Basel 1973).
142 CONSTANTINIDES, P.: Experimental atherosclerosis, pp. 55–56 (Elsevier, Amsterdam 1965).
143 CRAWFORD, M.D.; GARDNER, M.J., and MORRIS, T.N.: Mortality and hardness of local water supplies. Lancet *i:* 827–831 (1968).
144 BENTLEY, J.P.; WUTHRICH, R.C., and BUEREN, A.M. VAN: Lathyrism and mucopolysaccharide metabolism in aorta, skin and cartilage. Atherosclerosis *12:* 159–172 (1970).
145 LALICH, J.L.; PAIK, W.C.W., and ALLEN, J.R.: Production of arterial haemosiderosis in rhesus monkeys following the ingestion of β-aminoproprionitrile. Lab. Invest. *25:* 302–308 (1971).
146 ROSS, R. and KLEBANOFF, S.J.: The smooth muscle cell. J. Cell Biol. *50:* 159–171 (1971).
147 WICKS, M.P. and GARDNER, D.L.: Microchemical determination of enzyme activities in the lathyritic chick embryo aorta. Br. J. exp. Path. *54:* 422–428 (1973).
148 PETERS, T.J. and DUVE, C. DE: Lysosomes of the arterial wall. Expl molec. Path. *20:* 228–256 (1974).
149 ALLEN, J.R.; CARSTEN, L.A., and KNEZEVIC, A.L.: Crotalaria spectabilis intoxication in rhesus monkeys. Am. J. vet. Res. *26:* 753–757 (1965).
150 ALLEN, J.R. and CARSTENS, L.A.: Monocrotaline-induced Budd-Chiari syndrome in monkeys. Am. J. dig. Dis. *16:* 111–121 (1971).
151 GERMUTH, F.G., jr.: A comparative histologic and immunologic study in rabbits of induced hypersensitivity of the serum sickness type. J. exp. Med. *97:* 257–282 (1953).

152 LEVY, L.: A form of immunological atherosclerosis; in LUZIO and PAOLETTI Advances in experimental medicine and biology, vol. 1, p. 426 (Plenum Press, New York 1967).
153 WINKLE, M. VAN and LEVY, L.: Effect of removal of cholesterol diet upon serum sickness-cholesterol induced atherosclerosis. J. exp. Med. *128:* 497–513 (1968).
154 SCHWARTZ, C.J. and MITCHELL, J.R.A.: Cellular infiltration of the human arterial adventitia associated with atheromatous plaques. Circulation *26:* 73–78 (1962).
155 HOWARD, A.N.; PATELSKI, J.; BOWYER, D.E., and GRESHAM, G.A.: Atherosclerosis induced in hypercholesterolemic baboons by immunological injury. Atherosclerosis *14:* 17–29 (1971).
156 POSTON, R.N. and DAVIES, D.F.: Immunity and inflammation in the pathogenesis of atherosclerosis. A review. Atherosclerosis *19:* 353–367 (1974).
157 THOMPSON, J.E.: Production of severe atheroma in a transplanted heart. Lancet *ii:* 1088–1092 (1969).
158 PORTER, K.A.; THOMSON, W.B.; OWEN, K.; KENYON, J.R.; MOWBRAY, J.F., and PEART, W.S.: Obliterative vascular changes in four human kidney homotransplants. Br. med. J. *ii:* 639–651 (1963).
159 BENDITT, E.P and BENDITT, J.M.: Evidence for a monoclonal origin of human atherosclerotic plaques. Proc. natn. Acad. Sci. USA *70:* 1753–1756 (1973).
160 COX, G.E.; TRUEHEART, R.E.; KAPLAN, J., and TAYLOR, C.B.: Atherosclerosis in rhesus monkeys. IV. Repair of arterial injury. Archs Path. *76:* 166–176 (1963).
161 STAMLER, J.: The relationship of sex and gonadal hormones in atherosclerosis; in SANDLER and BOURNE Atherosclerosis and its origin, pp. 231–259 (Academic Press, New York 1963).
162 JAGANNATHAN, S.N.; MADHAVAN, T.V., and GOPALAN, C.: Effect of adrenaline on aortic structure and serum cholesterol in *Macaca radiata.* J. Atheroscler. Res. *4:* 335–345 (1964).
163 MEIER, R.N.; GREENHOOT, J.H.; SHONLEY, I.; GOODMAN, J.R., and PORTER, R.W.: Sex differences in the serum cholesterol response to stress in monkeys. Nature, Lond. 812–813 (1963).
164 MANN, G.V. and WHITE, H.S.: Influence of stress on plasma cholesterol levels. Metabolism *2:* 47–58 (1953).
165 GUNN, C.G.; FRIEDMAN, M., and BYERS, S.O.: Effect of chronic hypothalamic stimulation upon cholesterol induced atherosclerosis in the rabbit. J. clin. Invest. *39:* 1963–1972 (1960).
166 LORENZEN, I.B.: Experimental arteriosclerosis. Biochemical and morphological changes induced by adrenaline and thyroxine (Munksgaard, Copenhagen 1963).
167 BHATTACHARYA, S.K.; CHAKRAVARTI, R.N., and WAHL, P.L.: Noradrenaline induced myocardial infarction in monkeys given an atherogenic diet. Atherosclerosis *20:* 241–252 (1974).
168 LAPIN, B.E.: quoted by BOWDEN,. Folia primatol. *4:* 346–360 (1966) and personal commun.
169 CHISOLM, G.M.; GAINER, J.L.; STONER, G.E., and GAINER, J.V.: Plasma proteins, oxygen transport and atherosclerosis. Atherosclerosis *15:* 327–343 (1972).
170 GAINER, J.L. and CHISOLM, G.M.: Oxygen diffusion and atherosclerosis. Atherosclerosis *19:* 135–138 (1974).

171 MALINOW, M.R.; PERLEY, A., and MCLAUGHLIN, P.: The effect of pyridinol-carbamate on aortic and coronary atherosclerosis in squirrel monkeys *(Saimiri sciurea)*. J. Atheroscler. Res. *8:* 455–461 (1968).

172 MALINOW, M.R.; MCLAUGHLIN, P., and PERLEY, A.: The effects of pyridinol carbamate on induced atherosclerosis in cynomolgus monkeys *(Macaca irus)*. Atherosclerosis *15:* 31–36 (1972).

173 HARMAN, D.: Atherosclerosis, inhibiting action of an antihistemine drug chlorpheniramine. Circulation Res. *11:* 277–282 (1962).

174 HARMAN, D.: Effects of the antihistamine chlorpheniramine on atherogenesis and serum lipids. J. Atheroscler. Res. *10:* 77–84 (1969).

175 BAILLEY, J.M. and BUTLER, J.: The influence of antiinflammatory agents on experimental atherosclerosis. Nature, Lond. *212:* 731–732 (1966).

176 FRIEDMAN, M.; BYERS, S., and ST. GEORGE, S.: Cortisone and experimental atherosclerosis. Archs Path. *77:* 142–158 (1964).

177 MORRISON, L.M.; MURATA, K.: QUILLIGAN, J.J.; SCHJEIDE, O.A., and FREEMAN, L.: Prevention of atherosclerosis in sub-human primates by chondroitin sulphate A. Circulation Res. *19:* 358–363 (1966).

178 MORRISON, L.M. and BAJWA, G.S.: Absence of naturally occurring coronary atherosclerosis in squirrel monkeys *(Saimiri sciurea)* treated with chondroitin sulphate A. Experientia *28:* 1410–1411 (1972).

179 RUTENBERG, H.L. and SOLOFF, L.A.: Possible mechanism of egress of free cholesterol from the arterial wall. Nature, Lond. *230:* 123–125 (1971).

180 CONSTANTINIDES, P.; BOOTH, J., and CARLSON, G.; Production of advanced cholesterol atherosclerosis in the rabbit. Archs Path. *70:* 712–724 (1960).

181 GIERTSEN, J.C.: Atherosclerosis in an autopsy series. 10. Relation of nutritional state to atherosclerosis. Acta path. microbiol. scand. *67:* 305–322 (1966).

182 ARMSTRONG, M.L. and MEGAN, M.B.: Lipid deposition in atheromatous coronary arteries in rhesus monkeys after regression diets. Circulation Res. *30:* 675–680 (1972).

183 GOULD, R.G.; JONES, R.J., and WISSLER, R.W.: Lability of cholesterol in human atherosclerotic plaques. Circulation *20:* 967 (1959).

184 TUCKER, C.F.; CATSULIS, C.; STRONG, J.P., and EGGEN, D.A.: Regression of early cholesterol-induced aortic lesions in rhesus monkeys. Am. J. Path. *65:* 493–514 (1971).

185 FRIEDMAN, M. and BYERS, S.O.: Observations concerning the evolution of atherosclerosis in the rabbit after cessation of cholesterol feeding. Am. J. Path. *43:* 349–359 (1963).

186 HORLICK, L. and KATZ, L.N.: Retrogression of atherosclerotic lesions on cessation of cholesterol feeding in the chick. L. Lab. clin. Med. *34:* 1427–1442 (1949).

187 PETERSON, J.E. and HIRST, A.E.: Studies on the relation of diet cholesterol and atheroma in chickens. Circulation *3:* 116–119 (1951).

188 KJELDSEN, K.; ASTRUP, P., and WANSTRUP, J.: Reversal of rabbit atherosclerosis by hyperoxia. J. Atheroscler. Res. *10:* 173–178 (1969).

189 VESSELINOVITCH, D.; WISSLER, R.W.; FISHER-DZOGA, K.; HUGHES, R., and DUBIEN, L.: Regression of atherosclerosis in rabbits. Atherosclerosis *19:* 259–275 (1974).

190 DE PALMA, R.G.; HUBAY, C.A.; INSULL, W.; ROBINSON, A.V., and HARTMAN, P.H.: Progression and regression of experimental atherosclerosis. Surgery Gynec. Obstet. *131:* 633–647 (1970).

Subject Index

Absorption rates of fat 56
Acid, linoleic 31
–, mevalonic 54
Acidic mucopolysaccharides 63
Acylation in atheroma 59
Acyl-CoA: cholesteryl acyl transferase 62
Adipose tissue, oleic acid of 33
Adrenaline, inducing biochemical and morphological changes 78
Age changes 28
– – in elastic tissue 9
– determination factors 29
– of particular primates 29
Alloxan 68
– diabetic animals 69
Amino-oxidase 72
β-Amino-propionitrile 11, 73
– derived from the seeds of *L. odoratus* 72
Aneurysms, dissecting 10
– in an imported adult *Cebus albifrons* 24
Anti-elastin antibodies 9
Antigen-antibody complexes in the vessel wall 74
Aorta, atherosclerotic, in pigeon 59
– in coarctation 12
–, differences in the structure of 12
–, human 7, 8, 10
– involved by medial calcification 36
–, normal, from control animals 46
–, proximal, of the baboon 12
–, reactive contractile distal 12
Aortic lipase 81
– media, thickness of 8
Arcus senilis 34
Arterial intima, lipid accumulation in 58
– –, lipoprotein in 58
– perfusion 60

– –, cholesterol metabolism in 60
– –, transport through 31
Arteries of the squirrel monkey, changes in 26
Arteriosclerosis in chimpanzees, aortic and cerebral 57
Artery, injured by freezing 76
Aspirin and development of lesions 81
Atherogenesis, cholesterol-protein interaction as a factor in 51
–, haemodynamic factors in 64
–, human problems of 48
–, hypoxia as factor in 48
–, mucopolysaccharides in 63
–, phospholipid synthesis in relation to 63
–, theories of 4
Atheroma, acylation and transacylation in 59
–, cholesteryl oleate in 59
Atherosclerosis 3
–, aortic, and dietary level of protein 51
– in baboons 48
– in birds 1
–, cerebral 30
–, experimental, inhibition of 79
–, immunological injury in 74
–, male preponderance of 76
– in man 49
– – –, relationship of hypothyroidism to 69
– – –, risk factors of 42
– in marmoset 31
– in the Masai 1
–, naturally occurring 31
– in primates, spontaneous 24
–, proliferative 45
– in squirrel monkey, experimental pathology of 50

Subject Index

–, susceptibility of various species to 47
Atherosclerotic lesions in baboons 27
– –, regression of 81
– pigeon aortas 59
Autoallergic mechanism in the vessel wall 45

B_6-deficient rhesus monkeys 71
Baboons, atherosclerotic lesions in 27
– and cholesterol in atheroma formation 49
–, cholesterol level in 16
– – metabolism in 32
–, coronary artery of 28
–, depancreatised 68
– developing atherosclerosis under 'natural' conditions 48
–, electron micrographs of atherosclerotic lesions in 27
– fed on sucrose 67
– as a model 49
–, proximal aorta of 12
– and research in human atherogenesis 48
–, serum level of cholesterol in 32
– –, lipid in 32
Biochemical changes induced by adrenaline and thyroxine 78
Birds, atherosclerosis in 1
Blood platelets 75

Calcification, medial 11, 36, 71
Captivity, altering character and distribution of lesions 24
Carbohydrate, dietary 67
Carbon monoxide, dosage of 66
Cardiograms of macaques, differences in 15
Cebus albifrons 52, 53
– –, aneurysm in 24
– –, perfused isolated segments of aorta and coronary arteries from 60
– *apella* 53
– *fatuella* 71
Cebus monkeys, cholesterol-fed, coronary arteries of 52
– –, cholesterol metabolism in 32
Cercopithecus spp., pulmonary arteries of 36

Cerebral atherosclerosis in chimpanzees 30
– – in a gorilla 30
– vessels 30
Chemical changes in arteries during regression 83
Chimpanzee 57
–, aortic and cerebral arteriosclerosis in 57
–, cerebral atherosclerosis in 30
–, myocardial infarction in 34
–, thrombosed abdominal aorta in 34
Chlorpheniramine 80
Cholesterol 56
– acyl transferase; acyl-CoA 62
– in atheroma 59
– biosynthesis 54
– concentration and atherosclerosis in man 49
– – required before lesions 46
– ester hydrolase 81
– –, hydrolysis of 59
–, free, flux of 57, 82
– level in the baboon 16
– metabolism 32, 55
– – in the arterial wall 60
– – in baboons 32
– – phospholipid ratio 63
– pool 49
– protein interaction 51
– rates of esterification 61
– rise following cholesterol feeding 55
– synthesis by liver slices of squirrel monkeys 54
– synthesised by the arterial wall 59
Cholestyramine 84
Choline, lack of 71
Chondroitin sulphate 79, 81
– – in endothelial cells 63
Coagulation factors in non-human primate 35
Coarctation, studies of the aorta in 12
Cockerels, oestrogens in 77
Coconut oil 52, 55, 56
Collateral circulation 12
Conducting system 14
Confluent myocarditis in rhesus monkeys 35

Subject Index

Copper, deficiency and excess of 72
Corn oil 55
Corneal arcus 44
Coronary arteries 13, 27, 28
– – of adult cholesterol fed cebus monkeys 52
– –, distribution of 13
– –, lesions in the 25
– – of rhesus and vervet monkeys 27
– – of squirrel monkeys 51
– lesions in the chick, regression of 84
– – in howler monkeys 36
– – in the infants of Yemenite Jews 47
– occlusion 34
Crocetin 80
– increasing oxygen diffusion through the plasma 80
Crotalaria spectabilis 74

Desmosine 9
Determination of primate age 29
Diabetes mellitus induced by alloxan 68
Dietary carbohydrate 67
Diffuse intimal thickening 28
Distribution of vasa vasorum 56
DNA in rhesus monkey aortas 62
Dynamics of flow 13

Effects of vitamins 70
Egg yolk, infusion of 58
Elastic tissue, age changes in 9
– – fragmentation 47
Elastin, ratio of collagen to 12
– synthesised by smooth muscle cell 73
Elastolytic enzymes 10
Elastosis 10
Electrocardiogram, variations in 14
Electron micrographs of atherosclerotic lesions in baboons 27
Endothelial cells, hyaluronic acid in 63
– –, pinocytotic vesicles of 58
– damage leading to platelet accumulation 65
– permeability 81
– – increased by histamine and 5-hydroxytryptamine 75

– surface 17
Entry of lipoproteins into the vessel wall 57
Esterification of cholesterol, rates of 61
Evans blue 63
Excess of trace elements such as cadmium 72
Experimental animal, fatty streak in 2
– atherosclerosis, inhibition of 79
– hypertension 64

Fat, absorption rates of 56
Fatty acid composition of cholesteryl esters 53
– – – of lipoproteins 53
– – – of phospholipids 53
– – – of triglycerides 53
– streak 1–3
Fibrinogen levels in non-human primates 35
Fibrinolysis in non-human primate 35
Fibrous plaques 3
Flow and shear 19
Fructose 54

Gelatinous lesion 2
Glycocalyx on the cell 64
Gorilla, cerebral and spinal atherosclerosis 30
Growth hormone 13

Haemodynamic factors in atherogenesis 64
Hard and soft water areas 72
'Häutchen' preparations 17
Hepatic lipogenesis 53
– veins, occlusion of 74
High density lipoproteins 56
– – –, cholesterol content of 54
Housing methods, effects of 48
Howler monkey 24
– –, coronary lesions in 26, 36
– –, mediocalcinosis in 35
5-Hydroxytryptamine increasing endothelial permeability 75
– in platelets 75
Hyperglycaemia in relation to the degree of atherosclerosis 69
Hyperinsulinaemia in $M.\,mulatta$ 67

Subject Index

Hyperlipidaemia 73
Hyperoxia 80, 84
Hyperresponders 43
Hypertension 42, 43
–, effects of, on atherosclerosis in the squirrel monkey 68
– in the genesis of atherosclerosis in man 64
– in monkeys 30
Hypertriglyceridaemia 67, 70
Hypervitaminosis D 71
– D_3 36
Hyporesponder 43
Hypothyroidism, effects of, on atherosclerosis in the squirrel monkey 68
–, relationship of, to atherosclerosis in man 69
Hypoxia in atherogenesis 48
Hypoxic injury 65

^{131}I dosage 77
– triolein tolerance curves 56
Immature rhesus and vervet monkeys 27
Immunochemical techniques 56
Immunological injury as a factor in atherosclerosis 74
Inhibition of experimental atherosclerosis 79
Insulin deficiency, effects of, on atherosclerosis in the squirrel monkey 68
– mechanisms in the obese 67
Intercellular junctions 19
Interspecies variation among primates 70
Intimal oedema 3
Ion-binding capacity of the vessel wall 12
Isodesmosine levels 71

Kallidin factor 80
Kallikrein factor 80

Lactic dehydrogenase 66
Lagothrix lagothricha 52
Lamellar units 7, 8
– – in the aortic wall of mammals 8
– –, discrepancy in number and thickness 7
Lathyrus odoratus 10
– –, β-amino-propionitrile derived from the seeds of 72

LCAT 59, 61, 82
– activity in different species of animals 61
Lecithin cholesterol acyl transferase enzyme 55
Lipid accumulation in arterial intima 58
– –, phases in 58
– in the arterial wall in man 33
– in the eye 33
'– loading' effect of vitamin D_3 72
– metabolism 54
–, perfibrous 56
– in primate intima 58
– in the skin 33
– in tendons 33
Lipogenesis in monkeys 53
Lipogranulomatous arteritis 45
Lipoprotein, entry of, into the vessel wall 57
–, high density 31
– – –, cholesterol content of 54
–, low density 31
– peptide 58
α- to β-lipoprotein cholesterol, ratio of 47
β-Lipoprotein, synthesis of 61
–, toxic role of 54
–, uptake of 66
Lipoproteins in the arterial intima 58
α-Lipoproteins 16, 31
β-Lipoproteins 31
– cholesterol 16
Low density lipoprotein pool 57
– – lipoproteins 54, 56, 57, 73
Lymph-node permeability factor 80

Macaca fascicularis 53
– *irus* 46, 47
– –, fatty streaking in the aorta of 27
– *mulatta* 16
– –, myocardial infarction in 34
– *radiata* 56
Magnesium, deficiency and excess of 72
Male preponderance of atherosclerosis 76
Marmoset 31, 54
–, lipid metabolism in 54
Masai, atherosclerosis in 1
Medial arterial changes 35

Metabolism and synthesis of phospholipid 62
Mevalonic acid 54
– –, use of 60
Microfibrils 9
Mönckeberg's sclerosis 78
Monoclonal smooth muscle cells of atherosclerotic plaques 76
Monocrotaline 74
Morphological changes induced by adrenaline and thyroxine 78
Mucopolysaccharides 65, 72
–, acidic 63
– in atherogenesis 63
– level of the aortic wall, affected by thyroxine 79
– in the vessel wall 56
Myocardial infarction 34
– – in chimpanzee 34
– – in non-human primates 34
Myocarditis, confluent 35
Myointimal cells, necrosis of 46

Non-human primates, platelet counts in the blood of 35
– –, platelets in 35

Obesity 42
Oestrogen in cockerels 77
Oleic acid level of adipose tissue 33
Orang-utan, coronary occlusion of 34
Oxygen diffusion through the plasma 80

P wave, variation in 14
Papio hamadryas, effects of frustration on 79
– *ursinus* 53, 54
Peanut oil 45
Perfused isolated segments of aorta from *Cebus albifrons* 60
Perfusion of aortas of squirrel monkeys 60
–, arterial 60
Perifibrous lipid 56
Phenylbutazone 81
Phosphatidylcholine 58
Phospholipase A_2 31

Phospholipid to cholesterol, ratio of 62
– metabolism 62
– monolayers 59
– synthesis 62
– –, in relation to atherogenesis 63
Pig, pulmonary vessels in 37
Pigeon aortas, atherosclerotic 59
Plasminogen levels in non-human primates 35
Platelet accumulation caused by endothelial damage 65
– proteases 9
Polyunsaturated phosphatidyl choline 81
Proelastin 9
Proliferative atherosclerosis 45
Prosimian primates 12
Protein deficiency 72
–, dietary level of 51
Pulmonary arteries of *Cercopithecus* spp. 36
– –, thrombosis of 34
– collaterals, development of 13
Pyridinol carbonate 80
Pyridoxine-deficient diets, effects of 70

Rabbit atheroma, reversibility of 84
–, pulmonary vessels in 37
Regression 82
– of atherosclerotic lesions 81
– and chemical changes in arteries 83
– of coronary lesions in the chick 84
– diets and sequential changes in individual plaques 85
– of lesions 51
– – – in rhesus monkeys 82
Reversibility of rabbit atheroma 84
Rhesus monkey aortas, DNA in 62
– –, arterial lesions and serum lipid levels in 53
– –, B_6-deficient 71
– –, cutaneous xanthomata in 33
– – developing xanthomatosis 44
– – from India 35
– –, regression of lesions in 82
– –, vessels of 58
Risk factors of atherosclerosis in man 42
RNA factor 80

Saimiri sciureus 50, 52, 54
Scanning electron microscopy 17
Segments of aorta, perfused isolated, from *C. albifrons* 60
Sequential changes in individual plaques and regression diets 85
Serum sickness 74
SGOT in rhesus monkeys 16
Skin, lipid in 33
Smoking and coronary artery disease in man, association of 67
– as risk factor in atherosclerosis 42
Smooth muscle cells 3, 62
– – – of atherosclerotic plaques 76
– – – synthesising elastin 73
– – type 45
Social pressure 25
Sphingomyelin 58
Sphingomyelinase 31
Spider monkey 70
– –, arterial lesions and serum lipid levels in 53
Spontaneously occurring atherosclerosis in primates 24
Squalene 60
Squirrel monkey, cholesterol synthesis by liver slices of 54
– –, coronary arteries of 51
– –, experimental pathology of atherosclerosis in 50
– – fed on cholesterol 43, 51
– –, lipoprotein situation of 32
– –, perfusion of aortas of 60
– –, regression of primate atherosclerosis in 82
– –, *Saimiri sciureus* 50, 52
Stress 76
–, restraint equated to 77
– as risk factor of atherosclerosis 42
Sucrose 67
– in baboons 67
Susceptibility of various species to atherosclerosis 47
Synthesis of ester 60
– of β-lipoproteins 61

T wave, variations in 14
Tendons, lipid in 33
Thickening, diffuse intimal 28
Thrombi in the aorta of *M. mulatta* 34
Thrombosis of the main pulmonary artery 34
Thyroidectomy 43, 77
Thyroxine, administration of 77
–, affecting the mucopolysaccharide level of the aortic wall 79
–, inducing biochemical and morphological changes 78
Toxoplasmosis in New World monkeys 35
Trace elements, deficiency and excess of 72
– –, excess of 72
– –, lack of 72
Transacylation in atheroma 59
Transport through the arterial wall of lipids 31
– of cholesterol 55
Tropoelastin 9
Turbulence and shear on the intima 65
– – –, the role of 56

Uterine arteries 10

Vasa vasorum in the aorta 8
– –, distribution of 56
Vascular disease, hypertensive 11
Vervet monkeys 27
Vessel wall, autoallergic mechanism in 45
– –, calcification of 11
– –, entry of lipoproteins into the 57
– –, ion-binding capacity of 12
– –, mucopolysaccharide in 56
Vinyl casts 13
Vitamin B_6 deficiency, effects of 72
– D_3, 'lipid loading' effect of 72
Vitamins, effects of 70

Wandering pacemaker 14
Water areas, hard and soft 72
Wooly monkey – *Lagothrix lagothricha* 52

Xanthelasma 44
Xanthomatosis in rhesus monkeys 44

Yemenite Jews, coronary lesions in the infants of 47

THE LIBRARY
UNIVERSITY OF CALIFORNIA
San Francisco

THIS BOOK IS DUE ON THE LAST DATE STAMPED BELOW

Books returned on time are subject to fines according to the Library Lending Code. A renewal may be made on certain materials. For details consult ding Code.

| 14 DAY
NOV 2 76
RETUR
NOV 1 2
14 DA
MAR - 1
MAR 3 1977
14 DAY
SEP 1 4 198
RETURNED
SEP 1 5 1981 | **14 DAY**
JAN 1 2 1987
RETURNED
JAN 1 5 1987 | |

Series 4128